THE WASTE ISOLATION PILOT PLANT

A POTENTIAL SOLUTION FOR THE DISPOSAL OF TRANSURANIC WASTE

Committee on the Waste Isolation Pilot Plant

Board on Radioactive Waste Management

Commission on Geosciences, Environment, and Resources

National Research Council

NATIONAL ACADEMY PRESS
Washington, D.C.   1996

NOTICE: The project that is the subject of this report was approved by the Governing Board of the National Research Council, whose members are drawn from the councils of the National Academy of Sciences, the National Academy of Engineering, and the Institute of Medicine. The members of the committee responsible for the report were chosen for their special competencies and with regard for appropriate balance.

This report has been reviewed by a group other than the authors according to procedures approved by a Report Review Committee consisting of members of the National Academy of Sciences, the National Academy of Engineering, and the Institute of Medicine.

Support for this study on the Waste Isolation Pilot Plant was provided by the U.S. Department of Energy, under Grant No. DE-FC01-94EW54069. All opinions, findings, conclusions, and recommendations expressed herein are those of the authors and do not necessarily reflect the views of the Department of Energy.

Cover: Federal regulations require calculations to show that the Waste Isolation Pilot Plant (WIPP), if certified as a transuranic radioactive waste repository, is expected to isolate waste from the environment for the next ten millennia. Current plans call for the erection on site of permanent markers containing signs and symbols, intended as decipherable messages to warn future generations of the dangers to nature and to human health of digging into a filled and sealed repository below the surface.

Coincidentally, ten millennia is also the approximate age of the earliest known pottery from Asia. One millennium ago, a now-extinct Indian tribe, the Mimbres, lived in Arizona and New Mexico. The cover shows a Mimbres pottery design, perhaps representing the delicate balance of nature, using a man and two animals in a mobile arrangement. The Mimbres design is used by permission from *Art of a Vanished Race: The Mimbres Classic Black-On-White*, by Victor M. Giammattei and Nanci Greer Reichert, Published by Dillon Tyler, Publishers, P.O. Box 645, Calistoga, CA 94515.

The background photograph, provided by the Department of Energy Carlsbad Area Office, shows a close-up of a sample of Permian age salt crystals taken from the WIPP excavations. The permanence of the geologic salt formation (over 200 million years old) is an attractive feature of the WIPP site and illustrates the exceptional time scales of concern in nuclear repository design, time scales that extend well beyond the typical duration of most engineering projects, languages, and civilizations.

COMMITTEE ON THE WASTE ISOLATION PILOT PLANT

The National Academy of Sciences is a private, nonprofit, self-perpetuating society of distinguished scholars engaged in scientific and engineering, research, dedicated to the furtherance of science and technology and to their use for the general welfare. Upon the authority of the charter granted to it by the Congress in 1863, the Academy has a mandate that requires it to advise the federal government on scientific and technical matters. Dr. Bruce Alberts is president of the National Academy of Sciences.

The National Academy of Engineering was established in 1964, under the charter of the National Academy of Sciences, as a parallel organization of outstanding engineers. It is autonomous in its administration and in the selection of its members, sharing with the National Academy of Sciences the responsibility for advising the federal government. The National Academy of Engineering also sponsors engineering programs aimed at meeting national needs, encourages education and research, and recognizes the superior achievements of engineers. Dr. William A. Wulf is interim president of the National Academy of Engineering.

The Institute of Medicine was established in 1970 by the National Academy of Sciences to secure the services of eminent members of appropriate professions in the examination of policy matters pertaining to the health of the public. The Institute acts under the responsibility given to the National Academy of Sciences by its congressional charter to be an adviser to the federal government and, upon its own initiative, to identify issues of medical care, research, and education. Dr. Kenneth Shine is president of the Institute of Medicine.

The National Research Council was organized by the National Academy of Sciences in 1916 to associate the broad community of science and technology with the Academy's purposes of furthering knowledge and of advising the federal government. Functioning in accordance with general policies determined by the Academy, the Council has become the principal operating agency of both the National Academy of Sciences and the National Academy of Engineering in providing services to the government, the public, and the scientific and engineering communities. The Council is administered jointly by both Academies and the Institute of Medicine. Dr. Bruce Alberts and Dr. William A. Wulf are chairman and interim vice-chairman, respectively, of the National Research Council.

ACKNOWLEDGMENTS

The committee has spent countless hours over the more than ten years since its last full report on WIPP, in discussions with staff from DOE and its contractors, EPA, officials of the State of New Mexico, the Environmental Evaluation Group, community leaders from Carlsbad, and concerned citizens. In open meetings held several times a year for more than a decade, the committee has heard a wide diversity of views on WIPP.

Genuine concerns have been expressed without rancor or polemics, but with conviction and unfailing courtesy. This is a tribute to the community involved in and concerned about WIPP. For the committee, and particularly the chair, it has been a privilege to have participated in these discussions. We sincerely appreciate all of the information and insights gained and hope that our report will be of value in arriving at an appropriate decision on the proposal to establish a TRU waste site at WIPP.

The committee thanks the many anonymous reviewers who painstakingly read and criticized our report, which has benefitted considerably from their efforts.

Finally, we wish to express our appreciation to staff colleagues of the National Research Council's Board on Radioactive Waste Management, both past and present, who have done much to assist the committee in its task. Particular thanks are due to Tom Kiess, Angela Taylor, and Erika Williams, without whose efforts the report would not have been completed.

Charles Fairhurst, *Chair*
October 1996

CONTENTS

Executive Summary

The Waste Isolation Pilot Plant (WIPP) is a network of underground excavations at a depth of approximately 658 m (2,160 ft), in bedded salt formations near Carlsbad, New Mexico, in the southeastern corner of the state (Figure ES.1). WIPP is intended to serve as a permanent repository for transuranic (TRU) waste, which consists of a wide variety of materials (such as protective clothing, laboratory equipment, and machine parts) that have become contaminated with radioactive transuranic elements[1] during use in defense-related activities. These materials, from U.S. Department of Energy (DOE) facilities, currently are stored at several DOE locations around the country and are classified as either CH (contact handled) or RH (remote handled) waste.

DOE has been investigating the suitability of WIPP as a TRU waste repository since the 1970s and plans to submit an application to the U.S. Environmental Protection Agency (EPA) in the fall of 1996 for a certificate of compliance to open and operate the facility. To obtain the certificate of compliance, DOE must demonstrate that the WIPP facility will comply with relevant U.S. federal regulations—chiefly, the EPA's 40 CFR 191 and 40 CFR 194.

The National Research Council (NRC) Committee on the Waste Isolation Pilot Plant was formed in 1978 at the request of DOE to provide scientific and technical evaluations of DOE investigations at WIPP. The committee's statement of task charges it to report on the current state and progress of the scientific and technical issues that form the core of a submission by DOE to EPA for certification of the WIPP facility.

Because DOE's compliance certification application to the EPA consists largely of conclusions drawn from DOE investigations, it is timely to comment on results of committee evaluations and their implications with regard to the overall suitability of WIPP as a repository

for TRU waste (Box ES.1). This report presents these findings.

BOX ES.1

Uncertainty in Repository Performance

Assessing the performance of a radioactive waste repository over the long time periods of interest necessarily includes significant uncertainties. This is recognized by EPA in the standards for the disposal of TRU and high-level waste and spent nuclear fuel. Regarding the required degree of proof, EPA notes:

Performance assessments need not provide complete assurance that the requirements of Part 191.13(a) [i.e., the containment requirements] will be met. Because of the long time period involved and the nature of the events and processes of interest, there will inevitably be substantial uncertainties in projecting disposal system performance. Proof of the future performance of a disposal system is not to be had in the ordinary sense of the word in situations that deal with much shorter time frames. Instead, what is required is a reasonable expectation, on the basis of the record before the implementing agency, that compliance with Part 191.13(a) will be achieved. (40 CFR Part 194.13(b))

This review of WIPP should be read with this limitation in mind; the findings and judgments reached in this report could not be achieved with absolute certainty, but instead reflect a reasonable expectation for WIPP performance based on the available evidence.

Several general committee findings regarding TRU waste disposal at WIPP are worth noting. These findings are based on the characteristics of the waste and the salt medium and from scientific and technical studies at WIPP and at potential salt repositories in other countries.

• Although TRU waste contains long-lived radionuclides that require geologic isolation, the overall

---

[1]Transuranic elements are those elements with atomic number greater than that of uranium. Most are radioactive because of their emission of alpha particles. TRU waste contains those with half-lives greater than 20 years in concentrations exceeding 100 nanocuries per gram.

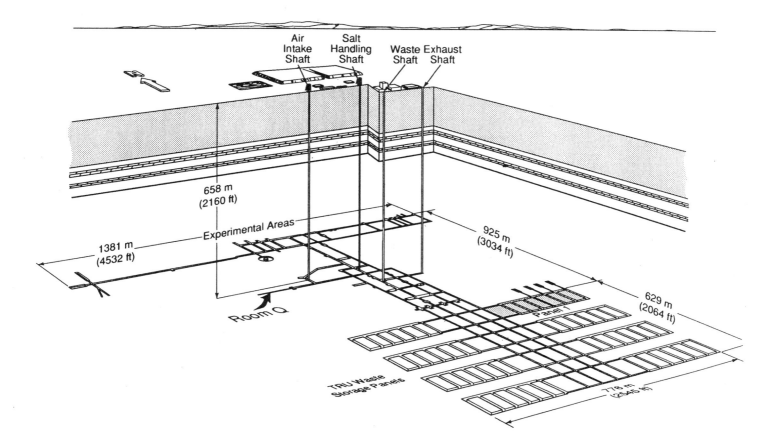

FIGURE ES.1 Three-dimensional view of the Waste Isolation Pilot Plant. The WIPP facility includes surface support buildings, a waste-handling building, four shafts, and the mined underground operations area. The repository is located approximately 658 m (2,160 ft) below the surface, within the Salado Formation, a Permian sequence of bedded salt with minor amounts of anhydrite and clay. The excavations are accessible from the surface by four vertical shafts. Only one of the planned eight panels, labeled Panel 1 in the diagram, has been excavated to date. Each panel consists of seven rectangular rooms, 10 m wide and 91 m long, separated by 30.5-m-wide pillars. Room Q, also labeled, is the site of a series of experiments on brine inflow into a 2.9-m-diameter, 109-m-long bored cylindrical tunnel. (Note: 1 meter [m] is approximately 3.28 feet [ft].) Source: Jensen et al. (1993), modified from their Figure 3-1.

level of radioactivity is much lower than that of high-level radioactive waste.[2]

• The early recognition of salt as an attractive medium for geological isolation (e.g., NRC, 1957) of radioactive waste has been confirmed by subsequent studies.

• Provided it is sealed effectively and remains undisturbed by human activity, the committee finds that the WIPP repository has the ability to isolate TRU waste for more than 10,000 years. The geologic stability and isolation capability of the Salado Formation, which consists of bedded salt, are the primary factors leading to this finding.

• The only known possibilities of serious release of radionuclides appear to be from poor seals or some form of future human activity that results in intrusion into the repository. The committee anticipates that the consequences of such human intrusion can be reduced based on available engineering design options and on improved understanding to be obtained from ongoing scientific studies.

• EPA's regulations (i.e., 40 CFR 191, as specified for WIPP in 40 CFR 194) relating to human intrusion do not take into account that if radionuclide releases to the environment via ground-water pathways at WIPP occur, they will be predominantly in non-potable water. This greatly reduces the risk of human exposure compared to a similar release in potable water.

GENERAL CONCLUSIONS AND RECOMMENDATIONS

The combination of general considerations, such as those outlined above, and detailed studies described later in this report, lead to the following conclusions and recommendations.

Based on available scientific evidence, the only probable threat to satisfactory isolation performance of the repository is the possibility of disturbance by human activity, deliberate or unintentional, that could compromise the integrity of the repository. Engineering methods are available, if needed, to reduce the consequences of human intrusion to acceptable levels.

Conclusion: Human exposure to radionuclide releases from transuranic waste disposed in WIPP is likely to be low compared to U.S. and international standards.

Consideration of the consequences of future activities that could violate the natural, or undisturbed, integrity of the repository is valuable for assessing the relative vulnerability of the repository to such activities and in identifying ways to reduce this vulnerability, but assessing human technologies thousands of years hence is highly conjectural and lacks a sound scientific foundation.[3]

Recommendation: Speculative scenarios of human intrusion should not be used as the sole or primary basis on which to judge the acceptability of WIPP (and, by extension, any geological repository).

Findings, Conclusions, and Recommendations on DOE Studies

DOE has concentrated on studies and calculations intended to determine compliance with federal regulations for WIPP in the event of human intrusion. Although the committee has not so restricted its studies, the following comments relate specifically to the DOE compliance activities. Most of the issues discussed below are significant only in the event of human intrusion.

1. *The role of Performance Assessment:* Performance assessment (PA) examines the combined effect of each component of the total system to assess the overall ability of the repository to isolate radionuclides from the biosphere. As used by DOE, PA responds to the question, 'Is WIPP in compliance with federal regulations?"

Although PA has made important contributions to the WIPP project, in retrospect, it is clear that the potential of PA is higher, and important opportunities

---

[2]The level of radioactivity per unit volume of WIPP TRU waste is of the order of 0.1 percent of the corresponding level for U.S. spent fuel [see Chapter 1 Table 1.1].

[3]Federal regulations (40 CFR 191, 40 CFR 194) require stylized calculations on releases due to human intrusion. The level of proof required is a "reasonable expectation." See Box ES.1. The weight to be given to human intrusion possibilities is also discussed in *Technical Basis for Yucca Mountain Standards* (NRC, 1995; see especially, pp. 11, 107-111,115).

to put PA to good use have been missed. PA is valuable at all stages of the repository evaluation process. It can identify the most critical components of the system, assess the significance of engineered supplements to the natural geological barriers, and serve as an aid to planning and management decisions on the most effective allocation of staff and project resources.

2. *Conservatism of performance assessment models:* The PA models currently used by DOE are too conservative in some respects. Such conservatism masks the potential for identifying and assessing the benefits of relatively simple engineering design procedures in reducing the consequences of human intrusion.

The assumption that the Disturbed Rock Zone (DRZ) bordering the room excavations remains a relatively high-permeability region throughout the first 10,000 years of the repository appears overly conservative. This is in marked contrast to the assumption (see Chapter 4) that the DRZ around the shaft will heal, achieving a permeability of between 10^{-16} m^2 and 10^{-18} m^2 within 50 to 100 years and approaching the essentially impermeable condition of intact salt within a small fraction of 10,000 years. Such a conservative assumption with respect to the DRZ in the PA models may prevent a realistic evaluation of the major benefits of compartmentation of the waste by room and panel seals in reducing the consequences of repository disturbance by intrusion.

3. *Complexity of performance assessment models:* DOE's PA modeling of radionuclide releases from WIPP involves complex combinations of many variables such that, to the non-specialist, it is not clear how the predicted releases depend on the component features, events, and processes in the geological isolation system.

The committee recommends that DOE develop, in parallel with the complex PA models, simpler versions that provide a more transparent, traceable path from the model inputs to the predicted releases. The insights gained from the simpler model as to which components of the isolation system are most critical to improved repository performance would serve a very useful role in decision making and in resource allocation for WIPP. It is essential, of course, that the simpler PA models still identify correctly the key features, events,

and processes upon which repository performance depends.

To illustrate this recommendation: because plutonium (Pu) is the dominant radioactive element of concern in the WIPP inventory, a simpler model could focus on Pu in the source term to the exclusion of other radioactive elements. However, understanding and predicting the behavior of Pu in the WIPP system is challenging, and experimental work with other actinides is necessary to develop the parameters for Pu required for the PA models, for both the full model and any simpler version. While studies of other actinides are necessary to support the chemical model developed for Pu, a simpler PA of the kind proposed here would consider only Pu isotopes as a source term, and only the dominant pathway(s) for environmental releases, with no more complexity than needed for an adequate representation.

4. *Waste characterization:* The waste characterization program being considered by DOE does not appear to be based on the needs for information important to an assessment of the long-term performance of the facility. Ideally, the PA should be used to determine what characterization is required.

5. *Nonradioactive constituents of TRU waste:* Nonradioactive hazardous constituents of TRU waste are considered to pose negligible long-term hazards compared to the radioactive constituents of WIPP waste (see Box 2.1 and discussion on 40 CFR 268 in Chapter 2).

6. *Behavior of salt at the WIPP site:* Time-dependent deformation of the salt and associated stratigraphic layers at WIPP is now understood well enough to allow reliable long-term calculations of salt deformation behavior as it relates to repository performance.

7. *Salado brine:* Small quantities of brine seep from the disturbed rock zone in the immediate vicinity of any excavation in the Salado Formation. The amount of brine accumulation is not sufficient to be a credible cause for significant escape of radionuclides from the sealed repository.

8. *Non-Salado brine:* Apart from possible effects of deep–well fluid injection in adjacent areas, brine flooding is only likely if, after loss of administrative controls, an intrusion borehole connects the repository with a deeper source of pressurized brine, such as has

been encountered by some deep boreholes in the vicinity of the repository.

9. *Gas generation in the repository*: Gas generation will be minimal in a dry or nearly dry repository such as WIPP because both chemical and biological gas-generating processes (e.g., metal corrosion and bacterial action on organic matter) require a liquid phase for mass transport of the reactants and products that are involved in gas formation.

10. *Treatment of waste:* Sophisticated treatment (e.g., incineration) of the TRU waste to be placed in a well-engineered WIPP repository is unwarranted to further improve repository performance, because gas generation is not a serious concern (see Chapter 3).

11. *Backfilling and compartmentation:* Simple repository engineering measures, such as backfilling of the rooms and tunnels in which the waste is emplaced, can be valuable and cost-effective methods of reducing the consequences of human intrusion and any associated brine flooding. Room and panel seals via backfill are relatively well-defined engineering procedures for improving the isolation process. In this regard, compartmentation is recommended by the committee to provide effective seals to eliminate hydrological communication between the waste-filled rooms.

12. *Non-Salado hydrology*: A more comprehensive understanding of the non-Salado hydrology is needed before a reasonable judgment can be made as to the role of the Rustler and adjacent formations in delaying radionuclide releases *in the event of human intrusion*. To date, studies have been overly focused on a two-dimensional analysis of the Culebra Dolomite. They have not sufficiently considered other possible hydrogeologic release pathways for radionuclides or interconnections between the Culebra and other formations. Potential releases to the Dewey Lake Red Beds, which are less conductive than the Culebra, but contain some potable water, are recommended for further study.

13. *Potash mining:* The consequences to the non-Salado hydrology of subsidence damage due to possible future mining of potash resources above the repository have not yet been evaluated by DOE. If the potential consequences are found to be seriously adverse, it is technically feasible to extract these resources pre-emptively, in a way that avoids subsidence and

associated effects and that reduces the potential for human intrusion by drilling.

14. *Deep well fluid injection:* The requirement to consider the effects on the repository of fluid injection activities was a relatively new addition to the final version of EPA criteria for certification (40 CFR part 194.32(c)). Neither the probability nor the effects on the repository from nearby injection of water or brine have been evaluated in detail by the committee, nor has DOE published an analysis of this issue. A comprehensive analysis of the risks and consequences of this scenario should be completed and documented.

15. *Waste solubility and transport:* The PA completed by DOE in 1992, and all subsequent analyses, consistently have identified a set of issues that will have the greatest impact on the compliance of WIPP *in the event of human intrusion* and associated brine flooding of the repository. Prominent among these issues are

- actinide solubilities in brine,
- formation and transport of colloids containing radionuclides, and
- retardation of radionuclides during transport through the Culebra.

The EPA also has identified these issues as critical to its evaluation of the compliance certification application.

At the time of the writing of this report, the data and models to be used to represent these three issues in the next version of the PA (to support the compliance certification application) were not available for review.

16. *Continuation of experiments and analyses:* Continuation of analyses and experiments recently initiated in the WIPP program to address concerns raised in issues 12, 14, and 15 is recommended by the committee, even though the results may not be available in time to be used in the compliance submission. Results of such testing could reduce uncertainties in the long-term performance of WIPP, eliminate concern over other issues, and be useful in judging the cost effectiveness of various waste isolation procedures at WIPP and other repositories.

SUMMARY

- Provided the WIPP repository is sealed effectively and undisturbed by human activity, the committee knows of no credible or probable scenario for release of radionuclides.

- For the WIPP repository disturbed by future human activity, the committee has noted three ways in which confidence in the performance of the repository could be increased:

1. Re-evaluation of the probability and/or consequences assigned to highly speculative scenarios of future human activities may reduce the estimated risk of radionuclide release.

2. Experimental and field programs in progress or planned may show that key parameters (e.g., actinide transport) are well within the range required to reduce the impacts of human activities on radionuclide releases substantially.

3. The implementation of available engineering options (e.g., compartmentation, treated backfill), which have not been considered in published DOE analyses, could reduce the consequences of human intrusion. The cost effectiveness of these options will depend on the outcome of (1) and (2) above.

The committee believes that some combination of the above three considerations will very probably be sufficient to allow DOE to demonstrate that a WIPP repository will keep radionuclide release within acceptable levels for the disturbed case.

Chapter 1

Introduction

The Waste Isolation Pilot Plant (WIPP) is an underground facility in bedded salt approximately 658 m (2,160 ft.) below the surface in a semi-arid region near Carlsbad, New Mexico, in the southeastern corner of the state (see Figure 1.1). A U.S. Department of Energy (DOE) facility, WIPP has been studied intensively to determine its suitability as a permanent repository for disposal of the category of intermediate-level radioactive waste known as defense-related transuranic (TRU) waste.

WIPP is a pioneering effort in the assessment of geological site suitability and design procedures for a waste repository. It is the first geological repository in the nation for which an application to begin permanent geological isolation is being submitted for a regulatory decision. If approved, WIPP will be the first "deep" designed repository in the world. (The low and intermediate-level radioactive waste repository at Olkiluoto, Finland, is located at a shallower depth of 125 m; the German repository near Morsleben is an abandoned salt mine, not designed initially as a repository.)

This report discusses the key technical issues that influence the suitability of WIPP for isolation of TRU waste. This chapter describes the radioactive wastes under consideration for storage at WIPP, the design concept of long-term storage in geological formations, and a brief description of the history of work at the WIPP site. Following chapters address specific technical issues that arise in considering the various pathways or scenarios by which radionuclides could move from the repository and be released to the accessible environment.

TRANSURANIC WASTE: WHAT IT IS, WHERE IT COMES FROM, WHERE IT MUST GO

Transuranic waste results chiefly from the production of nuclear weapons from plutonium and enriched uranium. The term transuranic indicates that the waste contains radionuclides with atomic numbers greater than 92, that is, greater than that of uranium (see Box 1.1). Existing amounts of the principal isotopes of these and other elements and estimates of their projected total quantity in DOE facilities are listed in Table 1.1.

TRU waste consists of a wide variety of contaminated materials from laboratory and production operations, including discarded protective clothing, laboratory test equipment and reagents, machine components, and solidified sludge. This waste has accumulated over the past 50 years as a result of weapons development and production at U.S. defense facilities. Future TRU waste generation is expected to come from cleanup of the contaminated sites and decommissioned facilities of the U.S. weapons complex. Packed in 55-gallon steel drums and wooden boxes, TRU waste currently is being stored at various sites across the nation.

Although not as hazardous as high-level waste (see Table 1.1), TRU waste contains long-lived radionuclides that, if released to the biosphere, pose a risk to humans and the environment for many thousands of years into the future. Of all the actinide isotopes, plutonium-239 (Pu-239) is the one of greatest concerns substantially beyond about 500 years because of its high concentration in the materials to be stored at WIPP (Table 1.1) and its long half-life. Because of its long-lived toxicity, TRU waste, like high-level waste, must be isolated from the biosphere for times greater than 10,000 years. Disposal in deep, stable geological formations, so that very little of its radioactive content will reach the accessible environment by any natural means, has been proposed. The WIPP facility has been designed for the disposal of defense-related TRU waste to meet these conditions.

BOX 1.1 What Is Transuranic Waste?

The Department of Energy, which is responsible for the management and disposal of defense-related TRU waste, uses the following definition: TRU waste is waste that is not high-level waste and that is "contaminated with alpha[1]-emitting radionuclides of atomic number greater than 92 and half-lives greater than 20 years in concentrations greater than 100 nanocuries per gram" (DOE Order 5820.2A). The Land Withdrawal Act of 1992 and the EPA 40 CFR part 191 contain the legal, regulatory definitions that agree with this language. Further details on exclusions and inclusions to WIPP-bound waste are contained in these references.

The most important transuranic elements for judging the suitability of WIPP as a TRU waste repository are plutonium (Pu), the major component; americium (Am), a moderate component; and neptunium (Np), a minor component because of its relatively lower inventory. Common usage of the term TRU waste often includes all elements in the actinide group with atomic number 90 (thorium) or higher. Thorium (Th) and uranium (U) isotopes are minor components of the WIPP inventory (Table 1.1), smaller components than americium. Other radioactive elements are present in the waste but in amounts so small that they have a less significant influence on the determination of the performance of the proposed repository.

Some of the isotopes of the actinide elements are long-lived alpha emitters with half-lives ranging up to billions of years. For many isotopes, however, the concentrations are low. Plutonium-239 (Pu-239), with a half-life of about 24,000 years, is the isotope of greatest abundance beyond about 500 years.

The radioactive constituents of TRU waste pose a radiological health hazard and are regulated by EPA in 40 CFR 191 and 40 CFR 194. Some TRU waste is mixed with chemically hazardous materials (e.g., certain toxic metals and organic compounds); the health consequences of exposure are derived from the chemical effect of these materials on the human body. EPA standards in 40 CFR 268 (also known as the Resource Conservation and Recovery Act of 1976, or RCRA) regulate the chemically hazardous constituents. An amendment (P.L. 104-201) to the WIPP Land Withdrawal Act (P.L. 102-579) exempts WIPP from federal RCRA requirements.

[1]Radionuclides that emit positively charged (alpha) particles in the process of decaying to more stable nuclides. Alpha radiation is the least penetrating of the three common forms (beta and gamma radiation are the other two) and cannot penetrate human skin. However, alpha emitters can be harmful if ingested or inhaled, or if they enter the body through other means, for example, through contact with a cut in the skin.

GEOLOGIC DISPOSAL OF RADIOACTIVE WASTE IN SALT

The objective of the U.S. nuclear waste disposal program is to place the waste in a location where harmful quantities cannot return to the biosphere by any foreseeable processes. The decision to develop a waste disposal facility in salt arose from an assessment by a committee of the National Research Council (NRC) appointed to study the problem of how to dispose of accumulating inventories of high-level radioactive waste. In its report, *The Disposal of Radioactive Waste in Land,* published in 1957, the committee recommended further work to assess the suitability of burying high-level radioactive wastes in stable geological formations at depths on the order of

500-1,000 m below the surface (NRC, 1957). Every other country that must deal with radioactive waste isolation, including Belgium, Canada, China, Finland, France, Germany, Japan, Russia, Sweden, Switzerland, Taiwan, and the United Kingdom, plans to use geological disposal to isolate radioactive waste.

The 1957 committee was particularly attracted to the notion of burying the waste in rock salt, either in salt domes or in thick salt beds It cited the following advantages:

• Salt can be mined easily.
• It is known to flow slowly under the pressure of overlying beds, and so will consolidate around the waste and isolate it in place.
• It is essentially impermeable.

TABLE 1.1 December 1995 Estimates of Inventory of Actinide Waste[a]

	CH-TRU	RH-TRU
Total anticipated activity (Ci), all isotopes	7,880,000	1,020,000
Design capacity of WIPP (m^3)	168,500	7,080
Total anticipated volume of WIPP (m^3)	110,000	27,000
Average activity WIPP TRU waste (Ci/m^3) [b]	46.8	143
Average activity of anticipated volume (Ci/m^3) [c]	71.6	37.8

Isotope	Half-life[e] (years)	Total CH-TRU (Ci)	Total RH-TRU (Ci)
Th-232[d]	1.41×10^{10}	9.11×10^{-1}	9.24×10^{-2}
U-233[d]	1.59×10^{5}	2×10^{3}	3.18×10^{1}
U-234[d]	2.45×10^{5}	5.53×10^{2}	3.93×10^{1}
U-235[d]	7.0×10^{8}	1.28×10^{1}	5.2×10^{0}
U-238[d]	4.47×10^{9}	3.96×10^{1}	1.44×10^{0}
Np-237	2.14×10^{6}	5.49×10^{1}	4.85×10^{-2}
Pu-238	8.77×10^{1}	3.8×10^{6}	1.45×10^{3}
Pu-239	2.4×10^{4}	7.82×10^{5}	1.03×10^{4}
Pu-240	6.5×10^{3}	2.08×10^{5}	5.07×10^{3}
Pu-241[d]	1.44×10^{1}	2.61×10^{6}	1.42×10^{5}
Pu-242	3.7×10^{5}	1.17×10^{3}	1.5×10^{-1}
Am-241	4.33×10^{2}	4.39×10^{5}	5.96×10^{3}

NOTE: CH-TRU = contact-handled transuranic waste; RH-TRU = remote-handled transuranic waste. The design capacity of RH-TRU is legally specified as 250,000 cubic feet (7,080 cubic meters). As the table shows, the anticipated volume specified in (DOE, 1995c) exceeds this quantity. This discrepancy could be resolved by a future change in the numerical estimates of anticipated volumes, or by a future renegotiation to augment the legally specified limit, or by waste processing to reduce the volume of the RH-TRU transported to WIPP to be less than the volume generated. Both DOE documents and EPA standards measure radioactivity in curies, where one curie, 3.7×10^{10} disintegrations per second, is the historical unit representing approximately the radioactivity of one gram of radium 226. The Systeme Internationale (SI) unit for radioactivity is the Becquerel, 1 disintegration per second. U. S. regulations express dose equivalent in roentgen-equivalent-man (rem) units, another historical unit differing from the SI unit, the Sievert (Sv), where one Sv is 100 rem. Because the U. S. work on WIPP uses these older units, they are used in this report.

[a] Not shown are minor concentrations of many other isotopes.

[b] The Ci/m^3 values shown here derive from the total activity (Ci) and the design capacity of the WIPP repository (m^3). By comparison, the average activity of U.S. spent fuel high level waste is 140×10^3 Ci/m^3 (10-year-old waste) and 140×10^2 Ci/m^3 (100-year-old waste). See Roddy et al. (1986).

[c] The Ci/m^3 values shown here indicate the ratio of total activity in curies (Ci) to the total anticipated volume (m^3) of waste to be shipped to WIPP.

[d] Because the half-life of Pu-241 is less than 20 years, it falls outside the EPA 40 CFR 191 definition of TRU waste, although it is a transuranic isotope. The thorium and uranium isoptopes are likewise actinides shown here that are excluded from classification as TRU waste, since they are not transuranic isotopes.

[e] The level of radioactivity decreases exponentially with time, with the half-life denoting the time at which the initial amount of any radionuclide is halved. After ten half lives, approximately 0.1 percent of the initial radioactivity remains. For long times, the WIPP inventory is dominated by Pu-239, ten half lives of which is a quarter of a million years. This time, at which approximately 0.1 percent of the initial amount of Pu-239 remains, is only roughly 0.1 percent of the present age of the geologic formation in which WIPP is excavated.

Source: (DOE, 1995c, Tables ES-3, ES-4, and 3-4).

• Salt that has existed underground for millions of years will almost certainly remain stable for thousands of years into the future.

These advantages, and others also, were noted in a subsequent study (NRC, 1970).

HISTORY OF WIPP

Selection of the present WIPP site came after rejection of a previous candidate site, an abandoned salt mine near Lyons, Kansas. Investigated from the mid-1960s to the early 1970s, the Kansas site was abandoned because of the large number of boreholes penetrating the formation and salt dissolution issues (Powers et al., 1978). Renewed efforts at a nationwide site selection and characterization process resulted in a focus of attention on the New Mexico region of the Delaware Basin (see Appendix A).

Exploratory drilling began in 1974 for a possible deep geologic TRU-waste disposal site in salt beds in New Mexico. The first location selected for study was rejected as the permanent site principally because of irregular subsurface geology (Powers et al., 1978). Recommendation of the present site was made in 1975. In 1976, the name "Waste Isolation Pilot Plant" was given to the project, and ERDA-9, the first exploratory hole located at the present site, was drilled (Powers et al. 1978; NRC, 1984). In 1979, the WIPP site, about 40 km, (25 miles), east of Carlsbad, NM in the southeastern corner of the state (Figure 1.1), was authorized, after preliminary testing, as a "research and development facility for demonstrating safe disposal of radioactive waste from defense activities and programs" (U.S. Congress, 1980). Construction of WIPP, together with studies of its suitability as a TRU repository, proceeded through the 1980s.

THE WIPP UNDERGROUND FACILITY TODAY

The WIPP facility is located approximately 658 m below the surface in the layered salt of the Permian Salado Formation (see Figure ES.1). Four vertical shafts from the surface provide access to the repository horizon and underground excavations. Drifts (underground roadways) link the shafts to the waste

disposal area, which is planned to consist of eight panels (the rectangular sections off the main haulage ways in Figure ES.1), each subdivided into seven rectangular rooms excavated between access drifts (corridors that provide passage to the individual disposal rooms, bounded by 30.5-m-wide pillars on each side). To date, only one of these panels has been excavated. Disposal room dimensions are 4 m high by 10 m wide by 91 m long. The drifts are essentially the same height but narrower than the rooms and are of varying lengths. Rooms have been excavated to follow the nonhalite marker beds that occur throughout the Salado, which dip at about 1° (from horizontal) to the south. Thus, the system of excavations is essentially horizontal.

THE DISPOSAL PLAN

The disposal plan involves stacking TRU waste-filled drums in each room until the room is filled. Some material may be packed into the spaces around the drums to fill the void spaces, a process termed "backfilling." Waste-filled rooms could be backfilled with crushed salt (excavated from the repository); crushed salt mixed with bentonite (a water-absorbing clay mineral); other materials combined with the salt (e.g., additives to reduce the solubility of plutonium in brine); or grout (Butcher, 1990; DOE, 1991). The specific plans for backfill have not been revealed in repository designs published prior to 1996.

A separate series of drifts and rooms for underground experiments has been excavated at the north end of the facility, several hundred meters from the disposal area (Figure ES.1). Various underground tests have been performed since the 1980s (see, for example, Butcher and Mendenhall, 1993), but these are now considered complete. The experimental area will not be used for waste disposal.

REGULATION AND LICENSING OF WIPP

In 1992, the WIPP Land Withdrawal Act (LWA; P.L. 102-579) transferred control of land at the WIPP site from the Secretary of the Interior to the Secretary of Energy and granted authority to the Secretary of Energy to close the area and its immediate surroundings to public use. A major provision of the LWA requires

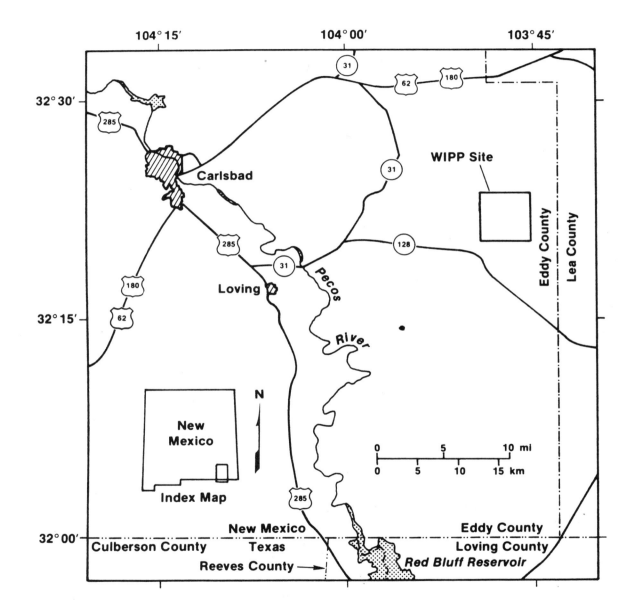

FIGURE 1.1 Location of the Waste Isolation Pilot Plant site. Inset map shows approximate location of map area in the state of New Mexico. Source: Modified slightly from Sewards et al. (1991, Figure III-1).

BOX 1.2 Compliance and Safety

The U.S. Environmental Protection Agency is responsible for ensuring that any U.S. geological repository for radioactive waste does not result in significantly adverse effects on human health and the natural environment. Federal standards have been promulgated to this end, and any proposed repository must be shown to be in compliance with these standards before it can be opened and operated to accept waste. When applied to a particular repository such as WIPP, features not anticipated in formulating the *compliance* standards could affect the overall *safety* of the repository. It is conceivable that a repository could comply with the standard, yet be significantly more or less safe than the measure of safety implicitly understood when the standard was formulated. This distinction could be important, because the federal standard (40 CFR 191) has never been applied to a real repository licensing situation.

The difference between compliance and safety can be illustrated by a simple example from the WIPP repository. EPA has defined the "accessible environment" as the region outside a 5-km-radius vertical cylinder around the center of the WIPP site, extending downward without limit from the surface (see Figure 1.2). DOE has chosen to define the boundary used in the compliance calculations as a smaller region, a 4 mile (N-S) by 4 mile (E-W) square region (6.4 km by 6.4 km square region) placed centrally within the 5-km-radius circle and also extending downwards without limit from the surface. Any migration of radionuclides across this boundary in excess of limits specified in the standard (40 CFR 191) constitutes a violation—that is, a failure to comply. This is independent of the depth at which the migration occurs and without consideration of the pathway by which the radionuclides would reach the surface.

Most of any "escaping" radionuclides would reach the accessible environment by transport in water. At WIPP, the most probable "escape route" to the surface is in the brine of the Culebra Dolomite, but this brine is for the most part undrinkable, even to cattle. This greatly reduces the safety risk compared to transport of the same quantity of radionuclides in fresh water. However, the containment requirement of the EPA standard does not address the potability of water containing the released radionuclides. Releases to salty, unpotable water near the surface thus may pose a *compliance* problem while presenting little *safety* hazard.

The committee found no probable situation for which the WIPP repository could be found to be in compliance with 40 CFR 191 and yet not be safe.

DOE to show that the repository will comply with federal regulations. The LWA designated the Environmental Protection Agency (EPA) as the regulator to determine whether or not compliance with appropriate regulations has been achieved (see Box 1.2). Chief among EPA regulations is the 40 CFR 191 standard, which describes general radiation protection requirements. Two other important regulations are EPA's 40 CFR 194, which describes the specific requirements at the WIPP site for compliance with 40 CFR 191; and EPA's 40 CFR 268 Resource Conservation and Recovery Act (RCRA) requirements on the chemically hazardous constituents of WIPP-bound waste.

Over the past two decades, DOE has conducted an extensive program of investigations to assess the suitability of WIPP as a TRU waste repository—that is, as a site that will adequately isolate the waste from the environment. Sandia National Laboratories has

served as chief technical adviser, with help from many sources, including the nearby Los Alamos National Laboratory, several other national laboratories, the U.S. Geological Survey, universities, and private engineering and consulting firms.

FRAMEWORK FOR THE REPORT

Presently, DOE is preparing an application to the EPA for a certification of compliance of the WIPP facility to demonstrate that the proposed repository meets regulatory requirements. Because the compliance certification application consists largely of conclusions drawn from DOE's scientific and technical investigations, it is timely to comment on results of the committee's investigations to date and their implications with regard to the overall performance of WIPP as a repository for TRU waste. Like the previous studies of the WIPP Committee (NRC, 1984), this report is a review of ongoing activities and should be viewed as a progress report rather than a final evaluation (see also Box 1.3).

Analyses of the performance and regulatory compliance of WIPP are addressed via performance assessment (PA), a process for identifying, for hypothetical future conditions, the potential pathways by which radionuclides could most probably be released to the environment. Chapter 2 reviews the allowable releases of radionuclides in EPA standards, describes the PA methodology used to demonstrate compliance with these release limits, and discusses PA results related to WIPP.

A detailed examination by DOE has been carried out over the past decade to study possible ways in which releases could occur, ranging from inadequate sealing of the repository to meteorite impact, climatic change, glacial erosion of the overlying formations to expose the waste, and many others. These possibilities were reduced by DOE to the 75 most credible features, events, and processes (FEPs), which were then examined in greater detail.

This examination indicated that eight combinations of events, or scenarios, warranted careful scrutiny (see Figure 2.2). From this group, the so-called "E1E2" scenario emerged as potentially the most serious way in which the isolation integrity of the repository could be violated. This scenario, which is discussed in more detail in Chapter 2 (see Figure 2.5), involves the drilling, some time within the next 10,000 years, of two oil or gas wells (either for exploration or for production), both of which intersect the repository. One of the wells also connects into a large reservoir of pressurized brine in the Castile Formation below the repository. It is further assumed that both wells would be "capped" (that is, sealed) near the surface and that the well casings would have corroded so that brine from the Castile could flow under high pressure through the repository, up the second hole, and into the Rustler Formation (principally the Culebra Dolomite), which lies above the Salado Formation. From there, the radionuclide-contaminated brine could flow more or less horizontally in the regional ground-water system, eventually reaching the accessible environment via a well drilled into the Rustler Formation to provide drinking water for cattle.

Detailed investigation of the factors that could influence the radionuclides released by the E1E2 and other scenarios has led to the identification of several issues that could be important in demonstrating the suitability of WIPP as a TRU waste repository. Chapters have been arranged to cover each of these issues. The performance assessment methodology used to calculate compliance integrates information from geological, hydrological, chemical, and repository design influences on radionuclide releases to the environment. The geological and hydrological influences of the salt beds that serve as host rock for the repository are discussed in Chapter 3. The significance of repository design options and waste treatment are discussed in Chapter 4. The chemistry of the radionuclides in brine and mechanisms of radionuclide

---

BOX 1.3 NRC Committee on WIPP

In March 1978, DOE asked the NRC to review the scientific and technical criteria and guidelines for designing, constructing, and operating the Waste Isolation Pilot Plant to isolate radioactive wastes from the biosphere. The panel (now committee) on the WIPP was established within what is now the NRC's Board on Radioactive Waste Management.

Since 1978, the committee has provided guidance to DOE on the scientific and technical adequacy of DOE programs designed to assess both the capability of the facility to isolate TRU waste and the overall performance of the facility. Two current members (Fred Ernsberger and former Chair Konrad Krauskopf) served on the original 1978 panel; three others (John Blomeke, Rodney Ewing, and Charles Fairhurst) began service in 1984.

The last comprehensive report issued by the committee was in 1984 (NRC, 1984). Eight letter reports on specific issues at WIPP (NRC, 1979a,b, 1987, 1988a, b, 1989, 1991, 1992) also have been produced by the committee.

The committee's current official statement of task is as follows:

> The Waste Isolation Pilot Plant Committee would prepare a report on the current status and progress of the scientific and technical issues that form the core of a submission by DOE to EPA for certification of the WIPP facility as a safe long-term repository for radioactive transuranic waste that complies with relevant regulations.

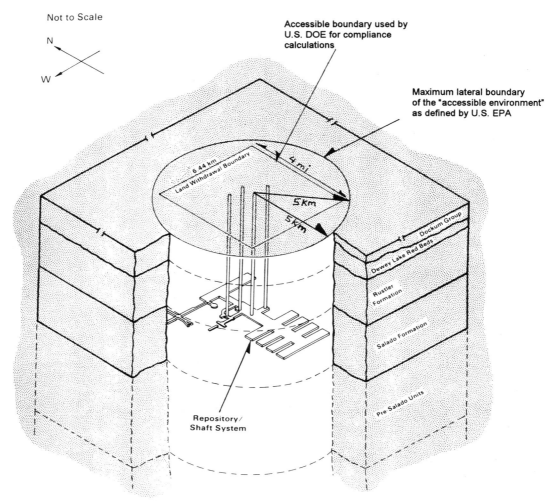

FIGURE 1.2   The EPA's "accessible environment" is shown here as the region outside a vertical cylinder of 5-km radius, centered on the WIPP repository.   DOE calculations are aimed at demonstrating compliance within a smaller region, a volume bounded by vertical sides that extend downwards without limit from a 4 mile by 4 mile (6.4 km by 6.4 km) square cross section.   Source: Adapted from Marietta et al. (1989, p. II-3).

transport in the subsurface environment are discussed in Chapter 5. Chapter 6 discusses the state of knowledge of the hydrological characteristics of the geological strata above the salt at WIPP, that is, the strata that would serve as a final barrier to releases of radionuclides to the accessible environment. Finally, Chapter 7 provides a summary of the conclusions reached in the committee's investigations. The first seven appendixes to the report provide supplementary information on specific aspects of the chapter discussions. An overview of the WIPP program, intended to assist readers in following the main themes and findings of the report, is presented in Appendix G.

The focus of the committee's review has been on the long-term, post-closure performance of WIPP. The issues associated with the operational phase, both surface and sub-surface, as well as the activities that must occur at other DOE sites, are not included in this evaluation. It is the committee's belief that the scientific basis for these other activities, both by DOE and by the various regulatory agencies involved, has been considered adequately in previous reports, except for a few issues that cut across both the operational phase and the post-closure phase. An example of this type of activity is waste characterization, which is discussed in Chapter 2.

Chapter 2

Regulatory Compliance and Repository Performance

To dispose of transuranic (TRU) wastes in the Waste Isolation Pilot Plant (WIPP), the Department of Energy (DOE) must demonstrate that the repository complies with applicable regulatory standards. These standards address

- transport of TRU wastes to WIPP,
- protection of workers from hazardous materials and mining hazards during operations, and
- the long-term performance of WIPP after wastes have been emplaced and the facility has been closed.

This report focuses on the third issue.

Two general standards of the Environmental Protection Agency (EPA), the first of which is defined in two parts, concern long-term repository performance:

    1.   a. EPA's *Standard for the Management and Disposal of Spent Nuclear Fuel, High-Level and Transuranic Wastes,* 40 CFR Part 191 (or 'Part 191'): 40 CFR 191 specifies requirements of geologic disposal systems in the United States.

        b. EPA's *Criteria for the Certification and Determination of the Waste Isolation Pilot Plant's Compliance with Environmental Standards for the Management and Disposal of Spent Nuclear Fuel, High-Level and Transuranic Radioactive Wastes,* 40 CFR Part 194 (or 'Part 194'): 40 CFR 194 addresses the specific application of 40 CFR 191 to WIPP.

    2.   The Resource Conservation and Recovery Act (RCRA) of 1976, 40 CFR Part 268: This law establishes a procedure to track and control hazardous wastes from the time of generation to the time of disposal.

This chapter is devoted mainly to a discussion of what is and what is not known by the committee about the long-term performance of WIPP in terms of the EPA standards. Compliance issues relating to RCRA are addressed in less detail. Issues associated with operations of the repository and transportation of waste to the repository are not considered. The most likely pathways, or "scenarios," through which radioactive materials could be released to the accessible environment are described, together with the performance assessment (PA) approach used to evaluate those scenarios against the EPA standards and some of the results to date. The chapter concludes with comments on the quality of the performance assessment work for WIPP, a brief discussion of other long-term radiological compliance issues, and some overall conclusions.

EPA STANDARDS FOR RADIOACTIVE WASTE

The EPA standard for long-term isolation of the types of radioactive wastes intended for emplacement in WIPP consists of both (1) the quantitative containment requirements and individual and ground-water protection requirements, and (2) the more qualitative assurance requirements, which are intended to provide confidence that the desired level of protection is achieved. The assurance requirements for waste characterization and monitoring are discussed later in this chapter.

A key aspect of the quantitative standards of 40 CFR Part 191 is the distinction between disturbed and undisturbed performance of the repository.

Undisturbed and Disturbed Repository Performance

Undisturbed performance refers to the case in which any releases of radionuclides to the accessible environment occur as the result of reasonably foreseeable natural processes. Releases due to human intrusion or from unlikely natural events—defined in

Part 191 as events having less than a 1 in 10,000 chance of occurring in 10,000 years—are excluded from the undisturbed case.

Undisturbed performance of the repository is governed by the individual and ground-water protection requirements of 40 CFR Part 191. The individual protection requirements limit the committed effective dose to any individual resulting from releases from the undisturbed repository to 0.15 milliSievert (15 mrem) per year. The ground-water protection requirements set concentration limits for potable sources of ground water that might come in contact with the wastes.

Disturbed performance includes consideration of inadvertent human intrusion into the repository. The EPA standard, Part 191, and supporting guidance require that releases of radionuclides resulting from drilling into the WIPP repository be considered, along with other specified disruptive events, named below, that could affect the performance of the repository. **The containment requirements, reproduced below, apply to the disturbed performance of the repository.**

Part 194 directs DOE to assume that the frequency of boreholes drilled into the WIPP site be based on the rate of drilling observed in the Delaware Basin during the 100 years prior to the time of the compliance application, by taking into account both deep drilling (i.e., boreholes that would reach the depth of the WIPP repository) and shallow drilling (i.e., boreholes that would not reach the depth of the repository). EPA specifies that the assumptions about future drilling practices with regard to borehole diameter and plugging practices should be consistent with practices in the Delaware Basin at the time of the compliance application.

Part 194 also requires that DOE consider the effect that mining of any potash of economic grade (in today's market) would have on the hydraulic conductivity of formations above the repository, and the effect that fluid injection for enhanced oil recovery in nearby oil and gas wells would have on the repository.

Containment Requirements

40 CFR 191.13 requires that

disposal systems for . . . transuranic radioactive wastes shall be designed to provide a reasonable expectation, based on performance assessments, that the cumulative releases of radionuclides to the accessible environment for 10,000 years after disposal from all significant processes and events that may affect the disposal system shall:

1. Have a likelihood of less than one chance in 10 of exceeding the quantities calculated according to Table 1 (Appendix A)[1]; and
2. Have a likelihood of less than one chance in 1,000 of exceeding ten times the quantities calculated according to Table 1 (Appendix A). (40 CFR 191, 1995, p. 11)

The key words in this regulation are "reasonable expectation," in recognition of the uncertainties inherent in making assessments over a 10,000-year time frame.

In paraphrasing 40 CFR 191.12, a performance assessment is defined as an analysis that identifies all significant processes and events that could affect the repository and evaluates the likelihood of each process or event and the effects of each on the release of radionuclides to the environment. To the extent practicable, these estimates are combined into an overall probability distribution displaying the likelihood that the amount of radioactive material released to the environment will exceed the values specified in Table 1, Appendix A to 40 CFR 191 (40 CFR 191, 1995, p. 11).

40 CFR 191 and 40 CFR 194 Radiation Dosage and Containment Requirements

In comparison to the radioactive waste standards adopted by most countries, Part 191 is unique in that, in addition to regulation based on radiation dose, repository compliance also is based on calculations of release fractions of selected radionuclides (Table 2.1). The containment requirement addresses the ability of a

---

[1]Table 1 of Appendix A of 40 CFR 191 is shown in this report as Table 2.1.

TABLE 2.1 Part 191 Containment Requirements for Selected Isotopes. Source: 40 CFR 191, Appendix A (1995)

Radionuclide	Release Limit (Ci/MTHM)
Thorium-230 or 232	10
Uranium-233, 234, 235, 236, or 238	100
Neptunium-237	100
Plutonium-238, 239, 240, or 242	100
Americium-241 or 243	100

The release limits specified here scale with the quantity of waste in a repository; for this reason, they are specified in terms of curies (Ci) that may be released per 10,000 years per 1,000 metric tons of heavy metal (MTHM). For a repository such as WIPP, which is intended to contain transuranic wastes, EPA has established in 40 CFR 191 that 1,000 MTHM is equivalent to 1,000,000 curies of TRU wastes with greater than 20-year half-lives. Therefore, the limits specified are applicable per million curies of TRU waste.

repository to isolate waste from the environment, without distinguishing releases that would lead to significant doses from those that would not. One advantage of this requirement is that, unlike a dose limit, it does not require assumptions about where people will live, how and where they will grow food, and what water sources will be used. A disadvantage is that the Part 191 containment requirements do not differentiate between transport of radionuclides off-site in a pristine aquifer that is a likely exposure pathway and—in the case of WIPP—transport via the Culebra Dolomite, in which the high content of dissolved solids makes the water unfit for human consumption.

Quantitative Assessment of Human Intrusion

Dealing quantitatively with human intrusion is a difficult issue. Any assessment of releases or risks from human intrusion into a repository is inherently arbitrary because the nature and rate of future intrusion over a 10,000-year time frame cannot be known. One advantage of basing the future human intrusion rate on the historical rate, as EPA does, is that the effect of known resources on past exploration and production activities is considered. However, whether or not this knowledge will predict accurately what may happen in the future—especially when institutional controls preventing intrusion have been lost—is uncertain. The difficulty does not concern specific details of EPA's guidance regarding how to treat intrusion in performance assessment, but rather that such guidance

is arbitrary and of unknown validity with respect to how future events actually will occur. The same issue is of concern in Europe and other countries dealing with the regulation of geological repositories. In general, radionuclide releases resulting from human intrusion tend to be treated qualitatively in these countries, for example, as a consideration in site selection.

The way in which human intrusion is handled in regulations 40 CFR 191 and 194 implies that WIPP, or any potential repository, could "pass or fail" on the basis of a human intrusion scenario that is, by necessity, arbitrary, especially in its assumptions concerning future technology. **Indeed, human intrusion rates or scenarios could be devised that would cause *any* geological repository to comply or fail to comply with EPA standards.** This issue of human intrusion has been considered at length in the NRC report *Technical Bases for Yucca Mountain Standards* (NRC, 1995).

40 CFR 268 Hazardous Waste Requirements

Until recently, federal law dictated that WIPP must comply with both the radioactive waste standard (40 CFR 191) and the RCRA standard (40 CFR 268), the latter of which establishes requirements for the disposal of chemically hazardous wastes (see Box 2.1). The 40 CFR 268 standard prohibits the land-based disposal of untreated hazardous wastes under most circumstances. This "land ban" has a "no-migration" provision by which a variance can be granted to permit the disposal

Box 2.1 Chemically Hazardous Waste Issues

Based on information presented by EPA and DOE contractor staff, it appears to the committee that RCRA performance will not be a limiting factor in the long-term performance of WIPP. Nonetheless, the following simple comparison can be made of the quantities and toxicities of the radioactive and chemical constituents of WIPP waste.

Source-term data reported in DOE's 1995 Draft Compliance Certification Application (DCCA) indicate that, in an average cubic meter (m^3) of contact-handled waste, the inventory of radioactive actinides with half-lives greater than 20 years, multiplied by EPA's risk factors for ingestion of these radionuclides, is equal to 8,500.* Thus, the risk from one cubic meter of contact-handled WIPP waste, if ingested, is quite high. By contrast, the inventory of volatile organic chemicals (VOCs) in 1 m^3 of WIPP waste, multiplied by the unit risk factors for inhalation of the VOCs, is reported in the WIPP Draft No-Migration Variance Petition to equal 46.

The EPA unit risk factors are based on daily inhalation of 20 m^3 of air for 70 years. Conversion of the risk factors associated with the inhalation of VOCs in 1 m^3 of waste gives a risk of 9×10^{-5}—that is, 9 chances in 100,000. This suggests that the relative toxicity (i.e., the ratio of the quantity of hazardous material, weighted by its toxicity) of ingested radionuclides to inhaled VOCs is of the order of 10^8. If the risk factors for *inhalation* (rather than for *ingestion*) of radionuclides are used, the toxicity of radionuclides relative to that of VOCs is of the order of 10^{10}.

This simple comparison of the quantity and toxicity of the waste components supports EPA's statement that the hazardous nature of WIPP waste is due to radioactive materials and that provisions for the long-term protection of public health and the environment from radionuclides in WIPP will also offer adequate protection against public health risks from chemical hazards.

* Normally, the product obtained by multiplying a quantity of hazardous material by its unit risk factor is the risk associated with the ingestion or inhalation of that material. However, a risk greater than 1 has no meaning. In such cases, the resultant product should simply be considered a toxicity-weighted inventory of hazardous material.

of untreated waste if the applicant can demonstrate that the waste will not move after disposal. In 1989, DOE submitted a No-Migration Variance Petition for the planned underground test program at WIPP and received approval for a 10-year period. In 1995, a Draft No-Migration Variance Petition was submitted by DOE for full-scale operations at WIPP.

During the 104th Congress, bills (U.S. Congress, House of Representatives, 1995, 1996; U.S. Congress, Senate, 1996) were introduced to amend the WIPP Land Withdrawal Act. Among other features, these bills would exempt WIPP from compliance with long-term RCRA requirements. In hearings on a House bill, EPA indicated that it was not opposed to such a change. An attachment to a letter from Mary Nichols, EPA Assistant Administrator for Air and Radiation, and Elliot Laws, EPA Assistant Administrator for Solid Waste and Emergency Response, to Senators Craig and Kempthorne, stated that

> . . . the Agency, therefore, believes that in the narrow context of the WIPP, which is subject to comprehensive regulation under the AEA [Atomic Energy Act], the WIPP LWA [Land Withdrawal Act], and RCRA, that a demonstration of no migration of hazardous constituents will not be necessary to adequately protect human health and the environment. (Nichols and Laws, 1995)

In this letter, EPA indicated that, under current law, it was obligated to continue to enforce RCRA provisions at WIPP. The letter illustrates the degree to which DOE and EPA have worked in a cooperative, constructive manner to address significant issues regarding compliance with RCRA regulations. On September 23, 1996, the President signed into law (P.L. 104-201) an amendment to the WIPP LWA that exempts WIPP from the federal RCRA requirements.

PERFORMANCE ASSESSMENT

Performance assessment (PA) encompasses the overall process of assessing whether or not a waste disposal system meets a set of performance criteria. For the WIPP PA, the system is a deep geologic repository disposal system in bedded salt for DOE TRU waste, and the performance criteria are various long-term environmental metrics in U.S. government regulations (Rechard, 1995, p. Glos-14).

Methodology

Iterative PAs of WIPP are being performed for DOE by Sandia National Laboratories. The performance assessments are intended to provide interim guidance while final compliance evaluations are being prepared.

The methodology has been developed to calculate a performance measure in the form of a complementary cumulative distribution function (CCDF) that permits comparison with the EPA release limits for radioactive waste disposal. The CCDF, a generally accepted form for depicting risk by a specific performance measure, was popularized by the Reactor Safety Study (U.S. Nuclear Regulatory Commission, 1975) for point estimates and by the Zion/Indian Point risk studies performed by industry (Pickard et al., 1981) for estimates with uncertainty. For WIPP performance assessments, the performance measure takes the form of normalized releases to the accessible environment of the radionuclides listed in Table 2.1. Under a normalized release limit, the estimated release is reported as the fraction of the release limit allowed under Table 1, Appendix A of Part 191. A release estimate equal to the allowable limit would be a normalized release of 1. Figure 2.1 is a hypothetical CCDF illustrating compliance with the 40 CFR 191.13 containment requirements.

The basic framework of the WIPP performance assessment methodology is the Kaplan and Garrick (1981) 'triplet" definition of risk. This definition of risk is founded on the principle that to determine risk, the following three basic questions must be answered:

1. What can go wrong?
2. How likely is it to go wrong?
3. What are the consequences?

'What can go wrong" is examined in the form of scenarios that can lead to releases of specific radionuclides to the accessible environment; 'how likely," is examined by calculating the probabilities of those scenarios; and the "consequences" are examined in terms of the overall likelihood of various levels of

release. The results are then cast into CCDFs of the form illustrated in Figure 2.1. The procedure for developing the CCDFs results in a structured set of scenarios leading to the specific consequences of interest; summing the probabilities of the scenarios for each release set; and plotting the complementary cumulative probabilities as a function of normalized release rates, usually on log-log coordinates (see Appendix B for more details).

A key part of the performance assessment model involves deciding on an appropriate set of scenarios, that is, identifying what can go wrong. The WIPP performance assessment team used the following five-step selection process for human-initiated scenarios (Cranwell et al., 1990):

1. compiling features, events, and processes (FEPs) that could affect the disposal system;
2. classifying events and processes to enhance consistency and completeness;
3. screening individual events and processes;
4. combining events and processes into specific scenarios; and
5. screening scenarios to identify and eliminate those that have little or no effect on the performance assessment.

Final selection of the scenarios in Figure 2.2 was based on screening events and processes according to probability, consequence, and physical reasonableness. The process resulted in the set of eight scenarios illustrated, although calculations were made on the basis of the first four scenarios only. A more detailed discussion of scenarios is presented later in this chapter.

The complexity in the performance assessment results from two main factors:

1. The numerous variables (approximately 50 in all) that represent the hydrological, chemical, mechanical, thermal, and transport processes involved in the transport and physical process models.
2. The effort required to perform sensitivity and uncertainty analyses on the models.

Documentation detailing the WIPP performance assessment effort includes a wide range of source materials documented in various reports and studies—including a 1992 Performance Assessment (the 1992 PA; Sandia National Laboratories, 1992, Vol. 3); the various iterations of the systems prioritization method (SPM; Sandia National Laboratories, 1995); and the

Draft Compliance Certification Application (DCCA; DOE, 1995a).

Uncertainties in the knowledge base were propagated through the scenarios, and sensitivity studies were performed so that results of the analysis include a quantification of the uncertainties associated with the calculated release rates.

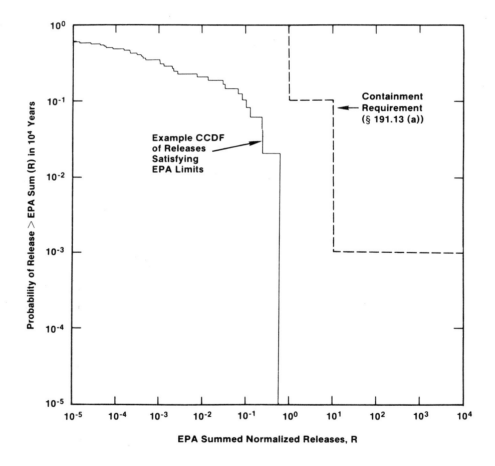

FIGURE 2.1 Hypothetical CCDF illustrating compliance with the containment requirements.   Source: Sandia National Laboratories (1992, Vol. 1, Figure 3.3).

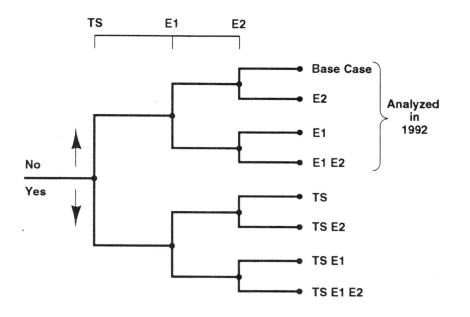

TS Is an Event in which Subsidence Results from
Mining of Potash

E1 Is an Event in which One or More Boreholes Pass
through Waste Panel and into a Brine Pocket

E2 Is an Event in which One or More Boreholes Pass
through Waste Panel without Penetration of a Brine Pocket

FIGURE 2.2 Potential scenarios for the WIPP disposal system. Source: Sandia National Laboratories (1992, Vol. 1, Figure 4.1). NOTE: TS refers to an event in which subsidence results from potash mining; the yes/no decision answers the question of whether potash mining is considered in the scenario. The base case depicted here represents the undisturbed repository. The E1 and E2 scenarios are discussed later in this chapter (see Figures 2.4 and 2.5). For further discussion of potash mining, see Chapter 4.

Results from Containment Requirement Calculations

On March 31, 1995, DOE submitted to EPA a draft analysis of WIPP compliance with 40 CFR 191 for the undisturbed case; this multivolume submission was supplemented in July 1995 with a performance analysis for the disturbed case (i.e., human intrusion; DOE, 1995a). DOE's current schedule calls for new performance assessment calculations for the compliance certification application to be completed and documented by September 1996, for subsequent submission to EPA a month later. Earlier performance results have been published in the 1992 PA (Sandia National Laboratories, 1992) and in the SPM analyses (Sandia National Laboratories, 1995). Compared to these earlier analyses, the DCCA included updates to the analytical codes and revisions of specific analytical assumptions, where dictated by EPA guidance (although this DOE submission predated the release of the final version of Part 194). The analytical assumptions of the DCCA appear to be largely unchanged from those of the 1992 PA and the SPM.

Although the DCCA indicates compliance with Part 191 containment requirements, the results of this analysis are too aggregated to allow any conclusions to be drawn about which scenarios and pathways contributed to the calculated releases. However, careful study of the 1992 PA does provide some insight into the relative magnitude of release scenarios.

CCDF curves summarizing radionuclide releases to the accessible environment resulting from cuttings removal and ground-water transport fall substantially below release limits promulgated by EPA. Although this is an important result, it must be noted that the rate of intrusion assumed in the 1992 analysis was less than is now required in the final Part 194, issued on February 9, 1996. The two main contributors to radionuclide releases to the accessible environment identified in the 1992 PA were (1) drill cuttings brought to the surface from exploratory drilling and (2) radionuclides transported by ground water flowing through the Culebra following human intrusion.

Some of the results from the 1992 PA with respect to the 10,000-year containment requirements of the EPA standard are as follows:

• Where the intrusion probabilities are based on expert judgment and dual-porosity transport with chemical retardation, the mean CCDF is *more than an order of magnitude below EPA limits*.[2]
• Where an intrusion rate of 30 boreholes per square kilometer per 10,000 years and dual-porosity transport without chemical retardation are assumed, the mean

CCDF is *approximately one order of magnitude below EPA limits*.
• Where an intrusion rate of 30 boreholes per square kilometer per 10,000 years and single-porosity, fracture-only transport (with little retardation) are assumed, the mean CCDF is *less than an order of magnitude below EPA limits*.

These three 1992 PA results represent a sample of all the different modeling assumptions and parameter values that were examined to assess their impact on the mean CCDF. The compliance outcome is a common result, with variations on the degree of sensitivity. Of course, individual scenarios can be found to produce CCDF curves that exceed the EPA limits, but for them the important issue is their likelihood, or frequency of occurrence (see Appendix B for more details). This information is used to determine the mean CCDF curve, the one used to assess compliance.

The containment requirement for WIPP (defined by Table 1 of Part 191, shown here as Table 2.1) specifies a release limit of 100 curies of plutonium per million curies of TRU waste, equivalent to a release fraction of about 10^{-4}, at least initially. Although release limits are specified in Part 191 for various isotopes, plutonium appears to be the most significant, in terms of both the inventory and compliance. Based on inventory estimates (DOE, 1995c) adjusted for the estimated date at which the repository would be closed, the WIPP inventory of thorium, uranium, neptunium, plutonium, and americium isotopes with half-lives greater than 20 years (which determine the release limit), is around 6 million curies. The corresponding release limit for plutonium-239 (Pu-239) would be 600 curies, or about 10 kg (approximately 20 pounds).

Figure 2.3, taken from the 1991 PA (Sandia National Laboratories, 1991), describes the radionuclide inventory as a function of time. The inventory is expressed in EPA units used for a single waste panel. An EPA unit is a convenient way to express the inventory, based on release limit definitions in Appendix A of 40 CFR 191. As can be seen, after about 500 years, Pu-239 is the dominant radionuclide of concern at WIPP.

---

[2]The CCDF results are in the form of mean values as permitted by Appendix C to 40 CFR 191. The means were calculated from a family of curves, each representing different percentiles, and are a direct result of the uncertainty analysis. A CCDF is generated for each "realization," which consists of a particular random choice of value of each of the model parameters (a choice within the range specified for that parameter). A mean CCDF is obtained from a set of CCDF curves generated this way, by averaging the cumulative release quantity for each value of the cumulative probability (see Appendix B).

Results Regarding Individual and Ground Water Protection Requirements

As noted earlier, the EPA standard sets forth two other quantitative requirements in addition to the containment requirements: (1) individual protection requirements; and (2) ground-water protection requirements. The individual protection requirement considers the radiation dose to humans in the accessible environment for 10,000 years of undisturbed performance. Because performance assessment results indicate that for the undisturbed case, brine (the medium for radionuclide transport) in contact with waste will not migrate more than a few "tens of meters" from the waste emplacement panels in 10,000 years, the

1992 PA did not include performance estimates for individual protection requirements. Subsequent assessments (e.g., SPM, DCCA) have also concluded that, under undisturbed conditions, waste would not migrate. The PA work (DOE, 1995a, Section 8.1) uses a bounding calculation to show doses to humans resulting from brine transport to be orders of magnitude below natural background; hence, compliance with the individual protection requirement can be demonstrated easily for the undisturbed scenario at WIPP.

The PA work has noted that the ground-water protection requirement does not apply to WIPP because there is no source of potable ground water as defined in the EPA standard (Sandia National Laboratories, 1992, Vol. 1).

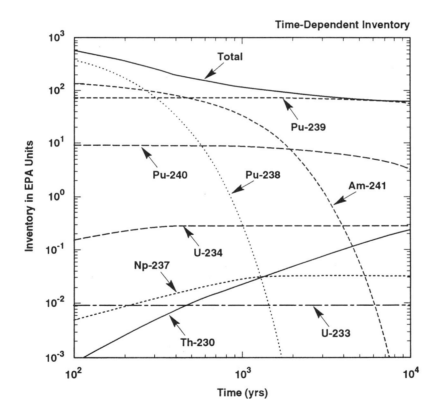

FIGURE 2.3 Time-dependent inventory expressed in EPA units for a single waste panel. Source: Sandia National Laboratories (1991, Vol. 4, Figure 2.4.2). More recent estimates (DOE, 1995c) of the WIPP inventory differ from those used in this figure.

In summary, the sole long-term radiological compliance issue is whether or not the containment requirements are met. Performance calculations indicate that if the shaft seals are effective, there will be little, if any, release or exposure from the undisturbed repository. For the disturbed case, compliance depends directly on the assumptions and analyses made for the frequency and consequences of human intrusion into the repository.

RADIONUCLIDE RELEASE SCENARIOS

Assessment of the likely performance of WIPP begins with the identification of scenarios by which wastes could be released from the repository. The scenarios that have been evaluated in greatest detail are those considered by DOE to represent the major mechanisms and routes for release of radioactive materials.

Undisturbed Case

Scenarios applicable to the undisturbed case considered in WIPP PA work include the following:

1. leakage of brines containing radioactive materials up the shaft and/or through a disturbed rock zone around the shaft up to the Culebra Dolomite and flow to the defined accessible environment via the Culebra; and
2. flow of WIPP brines directly along anhydrite marker beds in the Salado Formation-salt.

These scenarios are mentioned briefly here and represent mechanisms for potential release of radionuclides that underlie the considerations in Chapters 3 and 4.

Under the first scenario, human exposures could occur due to water extraction from the Culebra to add water to stock ponds,[3] and subsequent consumption of beef from cattle that have consumed this water. This

---

[3]Without dilution, ground water from the Culebra Dolomite immediately above the WIPP repository site is not potable by livestock or humans. Potable ground water from the Culebra Dolomite in the immediate vicinity is to the southwest, which is not on a projected pathline emanating from any hypothetical release.

scenario, involving flow out of the Culebra, is evaluated against both the containment requirements and the individual protection requirements.

The scenario involving flow along the anhydrite marker beds is evaluated against the containment requirements only, because the marker bed brines are not potable (Davies, 1989; Brinster, 1991; Yaron and Frenkel, 1994; Sexton, 1996), nor do they reach the surface.

Disturbed Case

Scenarios for the disturbed case involve releases resulting from boreholes drilled inadvertently into the waste. One scenario considered is the direct release of waste through drill cuttings brought to the surface. Other disturbed case scenarios include releases that occur through flow up a borehole into the Culebra dolomite, with subsequent transport in the Culebra; the effect of potash mining above the repository; and water flooding from nearby water injection for enhanced oil recovery.

Release of Waste Through Cuttings

The analysis of releases of radionuclides through cuttings brought to the surface in drilling mud includes several factors. In addition to the direct volume of waste displaced by the borehole (i.e., the area of the borehole times the repository height, with the assumption that all such wastes reach the surface), Sandia has made calculations of additional releases due to the erosion of wastes by drilling mud and the movement of wastes in the vicinity of the borehole because of internal gas pressure in the waste.

The quantity of waste intersected directly by a borehole can be determined simply from the frequency and diameter of boreholes assumed to be drilled for the period beginning 100 years after repository closure and continuing until 10,000 years after closure. The previous analyses based on the 1992 PA assumed 30 boreholes per square kilometer per 10,000 years; the current (1996) 40 CFR 194 requirement specifies that the historic record over the past 100 years should be the basis for the borehole rate, which, based on presentations to the committee by EPA and DOE staff,

Box 2.2   What Fraction of Waste Is Released Through Cuttings from Boreholes?

The degree to which cuttings may contribute to total releases in comparison to the release limits of Part 191 can be illustrated by the following simple calculation. Assume N boreholes per square kilometer per 10,000 years and the surface area of the repository occupied by waste **A** (m^2). The expected number of boreholes intersecting the waste over a 10,000-year period (I) is then given by

$$I = N \text{ (boreholes/km}^2) \cdot 10^{-6} \text{ (km}^2/\text{m}^2) \cdot A \text{ (m}^2).$$

The volume of waste W (m^3) displaced by these boreholes, by neglecting erosion and spalling and assuming a borehole area **B$_a$** (m^2), is

$$W = I \cdot B_a \cdot h,$$

where **h** is the repository height (m).

The fraction (F) of the repository that is released through these cuttings is simply the volume (W) of cuttings released, divided by the total repository volume (A·h):

$$F = W/A \cdot h.$$

Combining terms yields:

$$F = N \cdot B_a \cdot 10^{-6}.$$

leads to an intrusion rate closer to 45-50 boreholes per square kilometer during a 10,000-year period.

For 50 boreholes per square kilometer per 10,000 years with a 0.073-m^2 cross-sectional area (i.e., 12-inch diameter) borehole, the fraction F of waste released from cuttings (see Box 2.2) is 3.6×10^{-6} (this includes an adjustment to account for no intrusion during the first 100 years after closure). As noted above, the allowable fractional release is about 10^{-4}, initially, and increases with time. Thus, for 12-inch diameter boreholes drilled at a rate of 50 boreholes per square kilometer per 10,000 years, direct cutting releases apparently are less than 4 percent of the allowable limits. Even when allowing for increased release due to waste erosion by drilling mud and spallation, extraction of the waste as cuttings is unlikely to approach the release limits.

This calculation illustrates that the fraction of the Table 2.1 limit released as cuttings is independent of

1. the volume of waste disposed of in WIPP,
2. the geometry of the waste rooms (i.e., the ratio of height to width and length), and
3. the plutonium content of the waste.

To the extent that erosion and spallation are low in comparison to the volume of waste that would be released directly by displacement, the release fraction also does not depend on any waste treatment technology that may be employed. The cuttings release fraction *does* depend on (1) the assumed intrusion frequency, and (2) the assumed diameter of the intrusion borehole.

Therefore, the cuttings calculation is largely independent of the characteristics of the waste and the repository design. Furthermore, it is related to the location of the facility only to the extent that the location affects the assumptions made regarding the frequency and nature of intrusions over the next 10,000 years.

Release of Waste Through Boreholes: The E1 and E1E2 Scenarios

In the "E1" and "E1E2" scenarios, DOE postulates that a borehole through the repository may encounter a pressurized brine pocket in formations below the Salado Formation in which WIPP is located, resulting in the release of radioactive materials. The two specific scenarios analyzed for this type of release are illustrated in Figures 2.4 and 2.5. Figure 2.4 shows the E1 scenario, in which a single borehole penetrates the facility and continues downward into a pressurized brine pocket; Figure 2.5 shows the E1E2 scenario, in which two boreholes penetrate the repository. Because brine is assumed to migrate through the waste in the E1E2 scenario, this scenario is the more significant contributor to total calculated release.

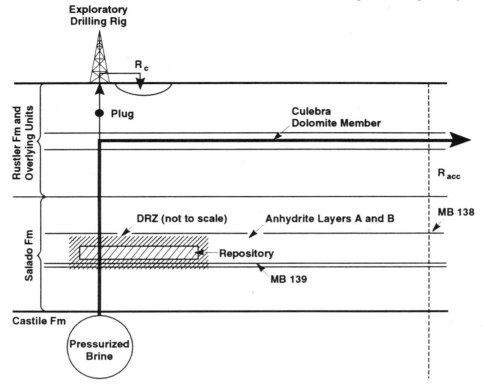

FIGURE 2.4 Conceptual model for the E1 scenario. Source: Sandia National Laboratories (1992, Vol. 1, Figure 4.2).

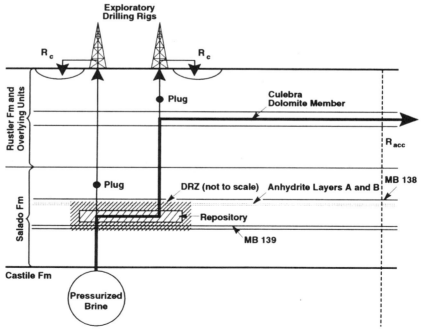

FIGURE 2.5 Conceptual model for the E1E2 scenario. Source: Sandia National Laboratories (1992 Vol. 1, Figure 4.3).

Such brine flow, presumably, would augment the salinity of existing water in the surface and shallow subsurface regions. As noted previously, ground water from the Culebra Dolomite at the WIPP site and in adjacent areas downstream of a hypothetical release already is too high in concentrations of dissolved solids to be potable to livestock or for humans. Regions of potable water from the Culebra are southwest of the site and are not in the direct path of a hypothetical release.

The rate and quantity of plutonium released to the accessible environment through the E1 and E1E2 scenarios depend on a number of factors. In PA sensitivity analyses, the three factors that were found to have a significant effect on calculated releases are

1. the solubility of plutonium in WIPP brines,
2. the potential for retardation of radionuclides in the Culebra, and
3. the potential for movement of actinides in colloids.

As a result of these PA results, expanded work to determine the effects of solubility, colloid formation and retardation on releases has been undertaken (see Chapter 5).

DISCUSSION OF PA MODELING EFFORTS

The committee believes that the PA model assessments of the E1, E1E2, and other scenarios involve technically unrealistic assumptions that have not been probed in past sensitivity analyses. For these scenarios, these technical assumptions also appear to be unreasonably conservative, specifically in the lack of credit given for compartmentation of the waste panels and rooms. These issues are discussed in more detail below.

Assumptions Regarding a Permanent Disturbed Rock Zone

Construction of the repository produces a network of stress-induced cracks that form a disturbed rock zone (DRZ) in the Salado salt bordering the excavated areas. The cracks are expected to close and heal in time due to salt creep. The permeability of the DRZ region, initially higher than that of the undisturbed halite because of these cracks, is restored to a value close to that of the intact salt over time.

The July 1995 update of the DCCA (DOE, 1995a) assumes that the salt around the repository between marker beds 138 and 139 (the anhydrite beds above and below the repository) remains an unconsolidated disturbed rock zone (DRZ) for the full 10,000-year regulatory period, with a permeability of 10^{-15} m^2. For comparison, the undisturbed halite is assumed to have a permeability that ranges between 10^{-24} m^2 (i.e., essentially impermeable) and 10^{-20} m^2. The draft compliance document does note that "although the DRZ is modeled conservatively in this assessment, it is the subject of a modeling study, and assumptions and treatment of this region may be different in the final Compliance Certification Application" (DOE, 1995a, p. 6-76).

The unconsolidated DRZ assumption appears to be inconsistent with the well-established properties of salt. A major feature of salt is that it behaves as a viscous liquid and creeps. The salt around waste rooms will reconsolidate as the pressure on the DRZ exerted by the backfilled rooms rises back to the lithostatic level. Such reconsolidation is estimated to take on the order of a few hundred years at most after repository closure.

A consequence of this conservative assumption is that boreholes that penetrate the salt between waste rooms are assumed to communicate with the waste and serve as release pathways, no matter when the boreholes are drilled. In addition, boreholes that penetrate *different panels* are analyzed as an E1E2 pathway, because under the conditions assumed throughout the entire repository, flow is calculated to occur via the DRZ. Sealing of rooms and panels so that this communication is prevented seems to the committee to be an entirely practical, cost-effective way to reduce the significance of the two-borehole (E1E2) scenario (see Chapter 4).

The more realistic treatment advocated here, of modeling time-dependent DRZ closure, adds complexity to the PA model. This complexity is warranted because it removes the non-physical conservative assumption that the DRZ does not heal. Removing an unrealistic assumption from the PA model has the added benefit of making the sensitivity analyses derived from the PA more accurate representations of the true dependencies on parameters in the model.

Characterization of Boreholes

It is conservatively assumed in the performance assessment that boreholes retain their initial diameter and have a permeability equivalent to that of a silty sand (10^{-14} m^2) for the entire 10,000-year regulatory period, based on present information of known and defensible natural processes that would fill the hole in time. No credit is given to mechanisms that would cause the boreholes to plug or shrink or that would interrupt the hypothetical flow of radioactively contaminated brine along the greater than 650-m borehole length. However, if further study should identify a natural mechanism that would cause the flow in the boreholes to be restricted, this could result in a significant reduction in calculated releases. Flow through an E1E2 scenario can occur only if both boreholes are open. With boreholes that stay open for times that are short in comparison to the 10,000-year regulatory period (e.g., 1,000 years), some borehole pairs from drilling events at random times will not result in a release because the first borehole will have closed before the second has been drilled. Further study to develop a better understanding of potential borehole closure mechanisms is recommended by the committee to examine an important component of the model calculation of releases.

Apparent Value of Compartmentation

As noted previously, calculations of releases over 10,000 years based on a DRZ that does not reconsolidate close to the value for intact salt are unrealistic (see also Box 4.1 in Chapter 4). This overly conservative assumption prevents the creep and healing behavior of the Salado halite from being properly included in past sensitivity analyses of WIPP performance. With effects such as DRZ reconsolidation excluded from a sensitivity analysis, an opportunity is lost to evaluate the full potential for cost-effective measures to demonstrate compliance. Stated another way, realistic DRZ modeling may lead to a demonstration of compliance using a cost-effective engineering feature such as backfill, which would

lessen the dependence of the compliance demonstration on the work addressing each of the three key factors enumerated above. These factors have emerged from the existing PA work and associated sensitivity analysis, and they involve technically complex experimental work. For example, addressing the first factor, that of plutonium solubility, involves completion of ongoing plutonium solubility tests that could be the time-limiting step in determining WIPP compliance, if no more straightforward and cost-effective way to demonstrate compliance were identified. If the DRZ healing behavior were included in the PA sensitivity analysis, confidence in compliance might be increased due to a lesser dependence on results from work in these three more complex areas.

Overly conservative (to the point of being physically unrealistic) analytical assumptions, such as those described above, not only lead to a pessimistic assessment of potential repository releases, but also prevent the identification of design factors that are important to performance. If one assumes, for example, that the salt reconsolidates in times much less than 10,000 years, then compartmentation of WIPP has enormous benefits regarding calculated releases through the E1E2 scenario. The probability of one borehole being drilled into any given area of the repository decreases in direct proportion to the area involved. The probability of two or more such boreholes penetrating the same waste area decreases even more rapidly as the area is reduced. When one also considers that, for flow to occur, boreholes must overlap both in location and in a time window determined by their duration, the great potential value of compartmentation is clear.

A high degree of compartmentation can be obtained at relatively low cost. The current plan calls for waste panels consisting of seven rooms plus access drifts along each end of the rooms in a 'ladder' arrangement (Figure ES.1). If, after the rooms are filled with waste, the access drifts are backfilled with compacted salt, then a high degree of isolation would be obtained. Changes in the design layout to ensure improved compartmentation seem entirely feasible, especially if, as is currently estimated (DOE, 1995c), the TRU waste inventory stored in WIPP is significantly less than the original design inventory. This issue is discussed in more detail in Chapter 4.

DISCUSSION OF REPOSITORY PERFORMANCE

The main mechanism through which individuals could be exposed to radioactive materials from WIPP is the migration (through mechanisms discussed below) of material carried by water that is to be used directly or indirectly by humans. Some ground-water sources at WIPP are too saline for consumption by humans or livestock or for irrigation, so therefore, these do not appear to pose a significant risk to humans, even if WIPP radionuclides were to migrate into them. The ground-water pathway that could most directly result in human exposure at WIPP is movement of water containing radionuclides into the Dewey Lake Red Beds (see Appendix A), in which existing ground water is known to be potable, with exposure resulting from direct or indirect human use of this water.

Although many analyses of radionuclide releases from WIPP have been made by Sandia National Laboratory and other DOE contractors, the radiation doses that could result have not been assessed for some reasonably identifiable scenarios and pathways. Most analyses of WIPP performance have focused on regulatory compliance with the EPA containment requirements, that is, calculations to estimate the quantity of radioactive materials that could be released across an arbitrary compliance boundary within 10,000 years. There has been less analysis of individual doses because the individual dose requirements of the EPA standard apply only to an undisturbed repository. For example, because of the high salinity of water in the Culebra, the analyses of releases to the Culebra do not directly translate into an assessment of doses to individuals. However, these analyses do contribute to understanding of how WIPP would perform because they include extensive consideration of processes, mechanisms, and parameters associated with the site and with the transport of radioactive materials from the repository.

Undisturbed Repository Performance

The individual protection requirements (i.e., EPA's 40 CFR 191.15(a)) require that the committed effective dose for the undisturbed repository not exceed 15 mrem/yr for 10,000 years. Because limitation of individual doses under the undisturbed case is a regulatory requirement, DOE has conducted analyses (DOE, 1995a; DOE, 1990; Lappin et al., 1989) of this issue. The process addressed by these analyses is the leakage of radionuclide-containing brines through shaft seals into the Culebra. Human doses are calculated based on the assumption that water from the Culebra dolomite is used in stockponds for cattle, and human exposure occurs through beef consumption (see Box 2.3). The individual doses for the undisturbed repository are estimated to have a peak value of about 10^{-8} mrem/yr, well below the EPA standard of 15 mrem/yr (and below the natural background level of roughly 300 mrem/yr).

Three important features of early analyses are that water in the Culebra appears to be too saline for consumption by cattle, transport by colloids was not considered, and pathways to the Dewey Lake were not included. In addition, the analytical treatment of the behavior of the shaft seals and repository in this analysis appears to be conservative, that is, likely to over-estimate the releases (see Chapter 4).[4] Taking all these considerations together, the committee concludes that the net effect will probably be to lower the already very low concentrations and doses indicated by DOE's analysis of radionuclide releases through shaft seals.

The ground-water protection requirements (i.e., EPA's 40 CFR 191.24(a)) require a calculation of levels of radioactivity in ground water in the accessible environment from an undisturbed repository for 10,000 years. These requirements were not evaluated in the 1992 PA work under the assumption that no relevant potable ground-water sources exist. Recent estimates of radioactivity released from an undisturbed WIPP repository to ground water are 10^{-3} pCi/l, well below not only the EPA standards (ranging from 5 to 15 pCi/l), but also the natural background level of roughly tens of pCi/l (DOE, 1995a).

Thus, for the effectively sealed and undisturbed WIPP repository, the committee has identified no

---

[4]The seal permeability of 10^{-16}m^2 results in no releases at 10,000 years (see Figure 4.1). As discussed in Chapter 4, the actual permeability is likely to be even lower than this design value, so that releases beyond 10,000 years are likely to be very small also.

Box 2.3   Human Exposure to Radionuclide Releases from WIPP

Human beings can be exposed to radiation in one of three ways: (1) direct external exposure, (2) exposure from inhaled materials, and (3) exposure from ingested materials. For most WIPP waste release scenarios, plutonium (Pu) is the main radionuclide of concern. Hence, the discussion here is specific to that element. The findings identified below for plutonium apply to other radionuclides in the projected WIPP inventory.

Direct external exposure to WIPP waste (via releases from drill cuttings brought to the surface) would not result in significant doses of radiation (DOE, 1995a), principally because plutonium's hazard is due to alpha radiation, which will not penetrate the skin.

Inhalation exposure occurs when radioactive material is inhaled and subsequently deposited in the respiratory tract or elsewhere in the body. Some low level of inhalation exposure could occur at WIPP if small radioactive particles in the drill cuttings at the surface were picked up and carried by the wind. In such a case, potential exposures to the drillers could be significant, but exposures to individuals off-site would be far less so (Rechard, 1995).

Ingestion exposure occurs when radioactive material that has been released into the accessible environment is ingested by a receptor (a human or some animal, plant, or water that may become part of the human food chain). As discussed in Chapters 5 and 6, the primary release pathway for radioactive material considered for WIPP is through the migration of plutonium in brine. Exposure by ingestion could occur at WIPP if radioactive material were transported underground in brines to a stock well, cattle drank contaminated water from the well, and humans then consumed meat from the contaminated livestock. However, the committee believes that the possibility of ingestion exposure occurring is remote, and that the potential health effects of this type of exposure are of little concern because:

1. Culebra water in the immediate vicinity of WIPP is not potable, even to cattle;
2. even if cattle were to drink the briny water, most of the plutonium would pass through unabsorbed, assuming that the gastrointestinal (GI) tracts of cattle and humans behave similarly (see point 4 below);
3. plutonium absorbed by livestock is deposited preferentially in bone, not meat; and
4. human ingestion is very inefficient in depositing plutonium in the body (a low absorption of Pu occurs in the human GI tract [EPA, 1988]—EPA guidance stipulates that only about one particle in 100,000 entering the GI tract would be absorbed, making ingested plutonium less toxic than inhaled plutonium).

The EPA standard regulates individual doses only for the case in which the repository remains undisturbed. Because the wastes do not appear to migrate under these conditions, DOE studies to date have not addressed specific health consequences of exposure to radiation from WIPP wastes in this case. Some preliminary DOE calculations of possible doses to individuals in the case of human intrusion by drilling suggest that the risk would be very low (see Rechard, 1995).

With the exception of the Dewey Lake aquifer, the major pathways by which plutonium and brine could migrate do not produce potable water. Certain concerns that have been raised by the committee about the possibility of transport of radioactive material to the Dewey Lake aquifer are discussed in Chapter 6. For a discussion of salinity limits appropriate for drinking water of livestock, see Yaron and Frenkel (1994, pp. 32-33, 42).

---

credible, probable mechanism for release of, or exposure to, radionuclides and concludes that DOE will be able to demonstrate compliance with the EPA standard by a wide margin.

Disturbed Repository Performance

As discussed earlier, DOE's PA showed that cuttings produced by drilling through the repository do not appear likely to cause the release limits to be exceeded. That is, although inhalation exposure from dust containing radionuclides that are brought to the surface as drill cuttings represented the dominant release scenario in the 1992 PA, the releases were within regulatory limits. If this is correct, compliance will depend on the results of other scenarios, such as the E1 and E1E2 scenarios, in which pressurized brine from below the repository horizon flows through the repository, into the Culebra Dolomite, and then into the defined accessible environment. Releases associated with these scenarios appear to have been calculated by using highly conservative assumptions about the disturbed rock zone around the repository.

Individual exposures that might result from WIPP in the event of human intrusion have not been analyzed extensively because such exposures are not covered by the standard. The scenario thought to represent the dominant exposure pathway is an ingestion exposure assumed to result from the following sequence of events. Two vertical boreholes are inadvertently drilled through the repository in the future, one of which punctures a hypothesized pocket of pressurized brine in the Castile Formation. As a result, brine from the Castile Formation flows through the repository, dissolving and entraining radionuclides that flow up the second drill-hole into the Culebra, which is above the repository but below the surface. These radionuclides would then sorb onto the mineral surfaces of the Culebra (see Appendix A for a more complete discussion of the geologic formations at the WIPP site). Ground water from this formation would then be withdrawn and fed to cattle[5], which subsequently, would be eaten by humans.

This Culebra-to-beef scenario is not the only one that is relevant to an assessment of all possible exposure pathways to humans. That is, using PA calculations of the Culebra pathway as a basis for inferring that minimal exposure would result from a WIPP repository depends on the assumption that there are no other pathways or scenarios worth considering. This is not yet certain. For example, in addition to the postulated main pathway through the Culebra, an alternative (or secondary) pathway exists through the shallower Dewey Lake Red Beds. The committee believes that the potential for individual doses received by drinking water drawn from this formation, located closer to the surface than the Culebra, should be analyzed and documented. That is, the first element of the triplet definition of risk, "what can go wrong," should include

the possibility of a borehole release from the repository to the Dewey Lake Red Beds. This is a concern because the Dewey Lake is known to contain potable ground water, and because it is the shallowest water-bearing unit in the area, it is also the easiest subsurface unit to be tapped for water supply by a future society. Thus, the potential for individual doses to be received by individuals (as a consequence of contamination from a breached repository) may be greater than those arising from an equivalent release into the Culebra. Because the Dewey Lake is less transmissive than the Culebra, the relative amount of plutonium reaching the Dewey Lake and the rate of lateral spreading may be less than for the Culebra. A quantitative analysis of this issue has not yet been presented to the committee in documented form for review.

Assessment of the isolation performance of the repository when disturbed by human intrusion is more complicated than in the undisturbed case. Because of gaps in the data available to date and the preliminary nature of the PA work reviewed by the committee to date (i.e., prior to the 1996 PA), it is not possible to make a compelling case that the radionuclide releases will comply with the EPA standard for the disturbed case. However, for the WIPP repository disturbed by future human activity, the committee has noted three ways in which on-going or additional work may lead to a demonstration of compliance. These are:

1. Re-evaluation of the probability and/or consequences assigned to highly speculative scenarios of future human activities may reduce the estimated risk of radionuclide release.

2. Experimental and field programs in progress or planned may show that key parameters (e.g., actinide solubility, sorption, radionuclide travel times, colloidal transport) are well within the range required to reduce the impacts of human activities such that releases fall within the acceptable range.

3. The implementation of available engineering options (e.g., compartmentation, treated backfill), which have not been considered in published DOE PA analyses, can substantially reduce the consequences of human intrusion.

---

[5]This Culebra-to-beef-to-humans pathway does not seem to account for the fact that the water withdrawn from the Culebra in areas downgradient from a hypothetical leak is too saline to be potable, even for cattle.

Based on the PA analyses published by DOE to date, which the committee believes include some very conservative assumptions (i.e., tending to over-estimate releases), the committee concludes that incorporating the above three considerations very probably will allow DOE to demonstrate that a WIPP repository can comply with the EPA standard for the disturbed case. These topics are treated in the remaining chapters of the report.

GENERAL QUALITY OF WIPP PERFORMANCE ASSESSMENT ACTIVITIES

Development of the WIPP PA has been a pioneering effort: it was the first detailed assessment of an actual radioactive waste repository to be published and has been the model for many subsequent analyses worldwide. The linkage of many computer codes and the capacity to use probability distributions for parameters and to propagate uncertainty through the analysis are major technical achievements. Nonetheless, in several important aspects, the full potential of the PA has not been realized in the WIPP project.

The committee noted several concerns above regarding the analytical treatment of some aspects of WIPP performance, for example, in the failure to consider salt reconsolidation and compartmentation of the waste rooms in the assessment. A more general problem is that the parameter values used in the PA reflect an uneven mixture of conservatism and realism. This is not unusual in risk assessment in which it is accepted practice to use a series of iterative assessments, based initially on conservative bounding values, followed by substitution of more realistic analyses that make a significant difference. A potential problem with this approach in the case of a complex analysis is that sensitivity analyses will fail to identify parameters that are most important to performance when those parameters are represented by a bounding value. A concern also is raised in Chapter 6 that some parameters assumed to be independent are, in fact, interrelated.

The PA identification of actinide solubility in brine and retardation as being among the most important parameters of WIPP performance is probably correct, but to a lesser degree than indicated by the sensitivity analysis. This is because other performance factors, such as DRZ permeability, were not sampled as analytical variables and, thus, their importance was not identified by the sensitivity analysis. As a result, an opportunity may have been missed to demonstrate compliance more quickly and economically than could be done through work on solubility and retardation.

In several iterations (the 1992 PA and SPM efforts), DOE has conducted sensitivity analyses of analytical parameters and used them to identify program research priorities. However, the PA modeling capability does not appear to have been used in a similar way to evaluate design alternatives, despite opportunities to do so. In 1991, DOE published the report of an Engineering Alternatives Task Force (EATF; DOE, 1991), a study conducted at least in part to respond to recommendations from the WIPP Committee (NRC, 1989). A similar study was published in September 1995 (DOE, 1995b) in response to EPA requirements. As described in Chapter 4, both of these reports included very positive comments about the value of sealing waste rooms; however, the current design (DOE, 1995b, p. ii) calls for panel seals but not room seals. Despite these two engineering studies, the PA model has not been used to evaluate what may well be the major potential benefit of such a design modification.

OTHER LONG-TERM RADIOLOGICAL COMPLIANCE ISSUES

Waste Characterization

Waste characterization—the process of identifying and classifying the chemical, physical, and radiological constituents of each drum of waste—is a critical aspect of every waste management project. For a project such as WIPP, many of the waste characterization needs are set by agencies such as the EPA, the New Mexico Environment Department (NMED), the U.S. Nuclear Regulatory Commission, and DOE itself. It is obviously impractical, if not impossible, simply to overlay the requirements of each of these individual agencies. Attempting to do so quickly degenerates into an exercise in a complete analysis of every possible component. Such analysis is costly and time-consuming and would result in the highest potential health hazard to the employees involved, with no clear need for all of the information collected.

DOE has in view the implementation of "performance-based" waste acceptance criteria for WIPP and extension of that concept to define waste characterization requirements (DOE, 1995a, Section 4.3.2, pp. 4-7). The concept is straightforward and reasonable: the waste characteristics that should receive the most attention in characterization are those that are most important from the viewpoint of the performance measures on which the isolation capability of WIPP is evaluated.

With respect to long-term radiological protection, the requirements proposed by EPA in 40 CFR 194.24(b) are consistent with a performance-based approach. For each performance measure important to isolation capability (e.g., activity, permeability, and porosity), EPA proposes that DOE use performance assessment to establish a range of waste characteristics within which the facility will comply with regulatory standards. For those waste characteristics that are comparatively unimportant to isolation, the ranges will be wide. This approach should help avoid unnecessarily detailed characterization efforts.

The extent to which existing waste drums will have to be opened and characterized is not yet clear, because specific details of the characterization required have not yet been established. In particular, the method by and extent to which a knowledge of historical processes can be used to characterize existing wastes have not been established. Similarly, the requirements for using statistical methods to characterize waste based on measured values from a subset of that waste are not yet defined.

In the committee's view, it is essential that any waste characterization program be established with a clear understanding of potential uses of the characterization information. Measurements of TRU waste characteristics can be expensive and may lead to occupational exposures. The value of information gained by such analysis must exceed the cost and risk of obtaining the information, but these trade-offs can be considered only if the sensitivity of repository performance to different waste characteristics is understood. Where costly waste characterization requirements appear to offer little value in terms of

ensuring health protection, DOE should seek to have the requirements reinterpreted.

Existing waste characterization regulations have been developed for surface waste management facilities rather than for a deep repository. These regulations reflect a general bias that more waste characterization is better than less and that the applicant has the burden of explaining why particular requirements should be waived in a particular case. 40 CFR 194 states that "any compliance application shall describe the chemical, radiological and physical composition of all existing waste proposed for disposal . . ." (40 CFR 194.24(a), Federal Register, p. 5240).

This is consistent with a performance-based waste characterization because a description of composition is not the same as a full characterization. The need for descriptive material is recognized. In the committee's opinion, there is no scientific basis for requiring a full characterization.

It appears that the project is concentrating too much on the problem of how to characterize waste and insufficiently on the question of whether to characterize it. The value of extensive characterization of WIPP wastes is questionable on several counts:

1. Whatever the current physical properties of wastes, they are likely to change significantly over the 10,000-year regulatory period of the facility. As the repository consolidates due to salt creep, the waste will be compressed and compacted.

2. The containment requirement of 40 CFR 191 limits the fraction of WIPP wastes that can be released over 10,000 years in terms of a percentage of the initial inventory; the absolute quantities that can be released are not defined. The total amount of radioactivity in WIPP will determine the release limits. For releases as cuttings, inventory information is largely irrelevant (see discussion of human intrusion earlier in this chapter). For releases due to solubility-limited flow, such releases may be largely independent of inventory, but release limits will depend on the total radiological inventory—that is, for the solubility-limited case, the more waste that is placed at WIPP, the easier it will be to comply with 40 CFR 191.

3. Performance assessments to date indicate that little, if any, waste migrates under undisturbed

conditions and that releases from human intrusion—assessed for various stylized scenarios, which include some apparently very conservative assumptions—appear to be small compared to the release limits. If the waste is not going to migrate to locations where humans may come into contact with it later, then the qualitative, common-sense answer to characterization is that it does not matter what the waste contains. A quantitative answer to the degree of characterization required would emerge rigorously from PA calculations, using the release limits specified by EPA standards, which apply irrespective of whether the releases pose a risk to humans.

Monitoring

DOE is required to perform a monitoring study as part of the information to be submitted in the compliance certification documentation. DOE is required to assess the feasibility of monitoring in both pre- and post-closure phases. A key element is that the monitoring techniques cannot impair or degrade the containment of waste.

In the committee's opinion, this DOE monitoring study should distinguish carefully between current technologies and expectations for future technologies. The committee believes that this study should also assess carefully those technologies that must penetrate the integrity of the closed repository to be effective. Simply stated, it may be easy to monitor parameters that will not provide any useful information about the facility and very difficult to monitor parameters that could be informative but almost assuredly require breaching the repository so that probes, cables, and other devices, can be routed to the surface. DOE and EPA must recognize that a monitoring program typical for RCRA facilities may be impossible to implement for the WIPP facility. Although the committee believes that the requirement for DOE to perform this monitoring study is appropriate, both EPA and DOE must be prepared to face a situation in which long-term, postclosure monitoring within the repository may be impractical. EPA, importantly, recognizes that such a monitoring program may be impossible to achieve without impairing the integrity of the repository. However, monitoring of the subsurface environment above the Salado (in particular, with sensors in the Culebra) could provide warning of a breach, probably without compromising the repository.

The 40 CFR 191 regulation require "long-term monitoring" of the closed facility, but without indicating what this implies. How long is long-term? What is to be monitored? The committee believes that long-term monitoring must be done from the surface. Observations that would require instrumentation inside the seals pose a risk of adding potential leakage pathways. The risk from monitoring systems that fail and become part of a leakage circuit could be higher than the risk due to leakage from a reliable unmonitored facility. Considerable advances are being made in remote investigation and monitoring by geophysical methods, but these have not yet matured sufficiently to provide monitoring of a repository. In summary, the committee believes that the EPA, in 40 CFR 194, has proposed monitoring requirements that do not appear feasible today.

The need for a monitoring study nonetheless provides an excellent example of how information from the operational phase at WIPP can be used to deal with compliance issues. It may be feasible, for example, to install probes and conduits into a room or panel that is then sealed. This would provide a direct measure of the ability to perform the monitoring anticipated in the regulations and to assess the reliability and integrity of the monitoring system. Should one of the seals fail because of the monitoring system, the consequences will be relatively slight because the failure could be recognized and repaired easily. A pilot study during the operating period should be informative regarding any early failures of the monitoring systems.

Peer Review and Quality Assurance of the WIPP Project

Two of the most common practices for enhancing the quality of technical work are a process of competent peer review and an appropriate quality assurance program. The WIPP performance assessment activity has been under substantial review, by peer review groups and others, since 1989. With WIPP, the question is not a lack of review, but rather the quality of the review and whether or not the reviews being performed are the ones needed. The reviews fall into three main categories: (1) formal technical reviews, (2) continuing review and feedback, and (3) miscellaneous review and advisory groups.

The first group of reviews is most likely to have continuity of impact. Reviews that appear to be particularly important are those from the PA Peer Review Panel, the State of New Mexico (both the Environmental Evaluation Group [EEG] and the Attorney General's Office), stakeholder reviews, and the EPA.

Despite the many peer review efforts and the superior technical ability of those performing them, the WIPP PA effort is centered on complex models, making meaningful probes difficult by outside review groups interested in understanding the effects and parameters most important to the outcome and interested in developing a simple, effective way in which these dependencies can be properly viewed. Given the earlier comments about the apparent conservatism in the analysis and the resultant misidentification of sensitivities, it appears that the peer reviews have overlooked these important shortcomings of the WIPP PA and have not resulted in a balanced, realistic assessment of WIPP performance.

One reason for this may be that the technical experts engaged in performance assessment have valued incremental technical innovations over transparency of the process. Understanding of the key aspects of WIPP radionuclide isolation performance need not be buried in a complex computer model. The descriptions of WIPP release scenarios described above are an attempt to indicate that the identified scenarios through which wastes could be released from WIPP are easy to visualize and that the factors contributing to releases can be identified readily. As the WIPP project moves forward in the compliance process, the hope of the committee is that the value of technical transparency and accessibility will be embraced by the project.

The quality assurance (QA) philosophy and procedures covering performance assessment are still under development. This documentation (see Sandia National Laboratories, 1991; 1992) covers QA procedures for (1) parameter selection and use of expert judgment, (2) analyses and report reviews, and (3) computer software supporting PA.

The development and implementation of QA procedures have progressed most visibly in the first of these three activities. Ideally, the QA program is to be implemented fully before final PA results are considered suitable for comparison with 40 CFR 191. However, full implementation alone is not sufficient to guarantee a high-quality result, which can be achieved only through the combination of a competent project team, peer review, and regulatory compliance. The final results have yet to be seen.

CONCLUSIONS

Performance assessment has a major role to play in demonstrating compliance with EPA standards and providing assurance to the public that transuranic waste can be stored without posing a significant health risk to the public in the Waste Isolation Pilot Plant. Although performance assessment did not originate with WIPP, the project has done much to advance the development and applicability of PA to radioactive waste disposal. As with any new development, be it a machine or a thought process such as performance assessment, initial expectations typically are much higher than actually can be achieved. Part of any criticism of the WIPP PA program may reflect these initial expectations.

Federal regulations relating to waste disposal define performance assessment as an assessment of the likelihood that the amount of radioactive material released to the environment will exceed a set of specified values. The analyses required to make such an assessment are capable of much more than merely demonstrating compliance with specific radionuclide release limits; they can be used on a much broader scale to enhance decision making and the overall efficiency of the WIPP project. The committee believes that these opportunities have not been exploited fully by DOE.

The general conclusion is that the WIPP performance assessment program is providing valuable insights on the performance capability of the repository, including the integration of features, events, and processes associated with total repository performance. The program is essential to success in opening the repository and can serve as the primary knowledge base for a meaningful risk management program during the operational phase. The extent to which performance assessment can become a primary monitor of repository performance clearly depends on the scope and resource base of the PA effort. The committee believes that scope must include performance measures that go beyond radionuclide release rates to the accessible environment. In particular, the WIPP PA, although quite responsive to the letter of the law, is too narrow in scope to serve as a robust basis for risk management— a goal beyond that explicitly required by federal regulations but clearly in the interest of public health and efficient management of public resources.

An example of an opportunity for continued use of PA in the manner recommended here is in waste characterization. PA can provide guidance to waste characterization requirements. The impacts, both in terms of economics and of human exposure to the radioactive materials, can be significant. However, both DOE and EPA seem committed to an extensive waste characterization program that is only minimally defined by the need for information as input to the PA process. EPA is apparently requiring waste characterization efforts well beyond that necessary for assessing performance of the closed repository. The

DOE seems intent on fulfilling these requirements with no clear scientific basis for the need for this information.

Remaining issues of concern relate to the timeliness of PA results, the transparency of the PA model, and its limited scope in terms of aiding WIPP program decision making. The complexity of the model and its limited set of performance measures handicap its usefulness in project decision making, because the mechanisms, processes, parameters, and events that contribute to risk are difficult to understand. If a simpler model could be developed that would represent the full model reasonably and adequately, such a simpler model—with expanded performance measures—would benefit the project. Similarly, the results and insights of the performance analysis, as reported in the DCCA, are largely inscrutable. A simplified description of the scenarios and associated processes under which waste from WIPP could be released and people could be exposed to radionuclides would be a valuable complement to the WIPP performance assessment report.

Chapter 3

Salado Hydrogeology, Gas Pressure, and Room Closure

Studies of physical, chemical, and biological processes involving the rock salt environment provide important conclusions about the behavior of waste-filled excavations in the Salado Formation. This information is reviewed briefly in the following four sections. The first discusses the hydrogeological properties of WIPP salt (see also Appendix C). The second draws general conclusions about gas generation. The third section discusses the influence of room closure, with local stratigraphy (Appendix A) and the creep behavior of salt (Appendix D) used as additional inputs. The final section discusses the combined effects of brine inflow, gas generation, and room closure.

SALADO HYDROGEOLOGY

In assessing deep geological disposal as a potential means of isolating radioactive waste from the biosphere and worthy of further research, the National Research Council suggested rock salt as a particularly suitable geological material (NRC, 1957). The fact that water-soluble formations such as the Salado Formation in New Mexico have remained in place for hundreds of millions of years indicates that, except for slow dissolution at the margins, they are hydrologically inactive. Another desirable characteristic of salt is that over time, it will flow around the waste and encapsulate it completely.

The Salado is approximately 600 m thick, extending about 400 m above and 200 m below the horizon selected for the WIPP repository. The Salado contains bedded, continuous layers of relatively pure halite, impure halite with clay and polyhalite constituents, and interbeds of anhydrite, accompanied by underlying clay seams (Freeze et al., 1995a, pp. 1-8).

Some of the interbeds are clearly evident and continuous over many kilometers and have been designated as "marker" beds. They serve as a guide in keeping the excavations within given bedded layers and

maintaining a fixed separation from the nearest anhydrite interbeds above and below.

Excavations at WIPP have not revealed any feature that would cause DOE to question the assumption of the very low permeability of the Salado. However, concerns have been raised (Bredehoeft, 1988) that although the Salado appears to be dry, it may still be saturated with brine, that is, the microscopic voids between the crystals that make up the salt beds could be filled with brine and the voids interconnected (see Box 3.1).

In this case, interstitial brine would tend to flow down the pressure gradient and into the excavated repository. Although the porosity of the halite is not high,[1] such a permeable-medium model considers the entire Salado Formation as a brine reservoir, able to flow into and fill the excavations. The brine would corrode the metallic waste containers, generating an underground "slurry" of contaminated brine. Through chemical and bacteriological processes, this brine might generate enough gas to raise the pressure above lithostatic (i.e., the pressure generated by the weight of overlying rock).

Several underground investigations and analyses of the interconnection model have been conducted, and a number of inconsistencies have been noted (Freeze and Christian-Frear, in preparation; Beauheim et al., 1991; McTigue, 1993). These include the following:

• Brine seepage into excavations in the Salado halite occurs in variable small quantities in unpredictable locations.
• Brine composition is quite variable, even among samples separated by distances of only tens of centimeters. Over a period of 230 million years (the

---

[1]The porosity of the Salado is generally about 1 percent by volume, and the interstitial moisture content is typically less than 0.5 percent by weight.

BOX 3.1 Permeability and Fluid Inclusions of the Salado Formation

The permeability of the Salado Formation salt is an important parameter in evaluating the capability of the WIPP repository to withstand hypothesized future flooding by an external source of water that would flow through any interconnected pores in the salt. Valuable insight into permeability is achieved by examining the composition of fluid inclusions. Fluid inclusions are tiny quantities of liquid or liquid plus gas enclosed in salt crystals of the Salado, which represent samples of the fluid from which the salt crystals formed or were subsequently recrystallized. Compositions of the inclusions show substantial variety, which has led some researchers (e.g., Roedder et al., 1987) to speculate that the salt may have recrystallized in the presence of different fluids that have flooded the salt beds at various times since their formation. However, more recent work (Bein et al., 1991) suggests that variations in chemistry of fluid inclusions is attributable to depositional and recrystallization processes that occurred millions of years ago and indicates that these variations do not require fluid flow through interconnected pores. The most recent work by Jones and Anderholm (1993, 1996) also suggests that the observed variations in fluid inclusion composition can be explained by natural and expected variations in the marine and marine-marginal environmental conditions during Salado deposition in the Permian. This means that the inclusion fluids probably have been isolated since the salt was formed. Thus, the Salado salt probably has remained essentially impermeable since its formation (see also Appendix C).

Salt mines elsewhere in the world have indeed flooded on occasion, but this is typically due to a combination of excessive extraction of salt during mining, presence of fractures in the salt caused by the disturbance of the mining, and close proximity to aquifers or surface water bodies. However, these conditions do not exist for the WIPP repository. The overburden pressure and deformation properties of salt (Appendix D) will close and heal any fractures that might develop in the disturbed rock zone and seal and isolate any residual gas-filled or brine-filled void spaces in the repository horizon. No permeable formations in the vicinity of the repository would serve as a suitable "source" or "sink" of ground water.

age of the Salado), these compositional differences would have been eliminated by diffusion had the pores of the formation been interconnected.

- Rock salt behaves essentially as a viscous fluid, with a zero yield (shear) stress, when the load is applied over a long time (see Appendix D). Such flow behavior would eliminate the possibility of connected pathways between the salt pores. Certainly, rock salt cannot sustain stress concentrations around microfissures over geological time. The inability of salt to sustain a shear stress over a long time is also indicated by the fact that the in-situ state of stress in the salt is isotropic (i.e., equal in all directions).

- Brine pore pressure in both halite and anhydrite marker bed units at WIPP has been found to be approximately 12 MegaPascals (MPa), which is between hydrostatic (6 MPa) and lithostatic (15 MPa). These measurements are consistent with isolated brine.

- In-situ testing (see Appendix C, and Beauheim et al., 1993b) indicates a large variability in intrinsic permeability, ranging from less than 10^{-21} m^2 for pure halite to as high as 10^{-18} m^2 (Freeze et al., 1995a) for anhydrite interbeds. When anhydrite is subjected to internal pressures (e.g., by high gas or brine pressures in the waste rooms) approaching lithostatic, its permeability may increase significantly due to the generation of incipient fractures and eventually open macroscopic fractures. Median values of 10^{-21} m^2 and 2.5×10^{-19} m^2 are used in computations by the Department of Energy (DOE, 1995a, pp. 6-73, 6-75) for halite and anhydrite permeabilities, respectively.

- Stress concentrations around underground boreholes and excavations in the Salado generate a disturbed rock zone (DRZ). The DRZ is a local region of enhanced permeability consisting of microfractures that spread at a slow and exponentially decreasing rate into the salt from any newly generated surface (created, for example, by borehole drilling).

All of these observations support an alternative to the interconnection (Darcy flow) model. In the alternative model, the undisturbed salt is essentially impermeable. Brine seepage comes from isolated domains of porosity containing brine that, in the undisturbed state, will be at lithostatic pressure. These domains become interconnected if they are penetrated by the micro and macrofissures of the DRZ. If the DRZ should extend to the marker beds, Darcy flow from these beds into the excavations could occur.

Appendix C addresses brine flow into excavations in the Salado and considers the following two possibilities: (1) impermeable salt containing permeable marker beds, from which brine flows into the excavations; and (2) a permeable Salado Formation with Darcy flow into the excavations. In both cases, by using the permeability values considered to be most appropriate on the basis of current data, the total inflow appears to the committee to be too small to result in significant brine accumulation in the rooms. With appropriate design width and sealing of the entrance drifts to the rooms, it should be possible to minimize extension of the DRZ to the marker beds and avoid connection between the beds and the excavation (see discussion of room seals in Chapter 4).

In conclusion, brine inflow to excavations in the Salado is likely to be significantly less than was thought to be the case several years ago—and will occur primarily via anhydrite marker beds. Careful attention to room filling and backfilling could aid in further limiting brine inflow.

GAS PRESSURE

Because the current design of WIPP calls for sealing all shaft communication between the repository horizon and the surface, gas generation could conceivably create a pressurized repository. Three sources of gas generation in the sealed repository have been identified (Lappin and Hunter, 1989):

1. radiolysis;

2. anoxic corrosion, caused by any inflow of brine that might react chemically with the carbon steel waste drums, generating gas (predominantly hydrogen); and

3. bacterial reactions with some of the organic constituents (e.g., cellulose).

The worst case scenario would involve total dissolution of all steel drums in brine, with generation of the equivalent amount of hydrogen, plus bacteriological conversion of all cellulosics, plastics, and rubbers to a mixture of methane, CO, and CO_2.

Based on this scenario, by considering gas compressibility but not solubility, reactivity, or the likelihood of interbed fracturing at lithostatic pressure, an early calculation (see Molecke, 1979a, b) indicated that the resulting pressure in the sealed repository could exceed lithostatic.

The generation of such quantities of gas cannot occur under undisturbed conditions because the quantity of brine required in this calculation is not available inside a sealed repository, and an external brine reservoir is sealed off from contact with the waste. However, the E1 and E1E2 scenarios do postulate an unlikely but possible mechanism for the introduction of very large quantities of pressurized brine. Accordingly, theoretical and experimental studies of gas generation have been undertaken (Freeze et al., 1995a) to provide estimates of the amount of gas generation to be expected under any credible set of conditions.

Early in the gas generation studies, DOE-sponsored researchers concluded that the amount of ionizing radiation in the expected WIPP inventory would not be enough to generate significant quantities of gas by radiolysis (see Molecke, 1979a, b). The remaining two sources of gas have been studied in more detail.

Laboratory Studies

Early laboratory experiments at Sandia National Laboratories (Molecke, 1979a, b) designed to permit estimates of chemical and biological gas generation rates were inconclusive because of the small and sometimes inconsistent effects observed. After a hiatus of several years, a larger program (Brush, 1990) was undertaken. This program, performed by recognized specialists, is now approaching completion (Brush, 1994; Beauheim et al., 1995).

The following conclusions about the major corrosion process have been reported:

Anoxic corrosion of steels . . . will produce significant quantities of H_2 and consume significant quantities of H_2O if

(1) . . . sufficient brine enters the repository after filling and sealing;

(2) significant microbial activity . . . does not occur (microbial activity will produce CO_2 or CO_2 and H_2S, which will passivate steels . . .) (Brush, 1994, p. 6),

and

Under humid conditions (gaseous, but not aqueous, H_2O present), anoxic corrosion of steels and other Fe-base alloys will not occur . . . (Brush, 1994, p. 7).

With respect to microbial activity, the conclusion announced in the same document is less definite: *". . . although significant microbial gas production is possible, it is by no means certain"* (Brush, 1994, p. 7).

Results of experimental studies (Brush, 1994) show that anaerobic gas generation rates in brine are likely to be quite low—approximately one micron per year rate of dissolution of the steel surface (Brush, 1994, p. E-9) and to be negligible if the waste is exposed only to gaseous phase "brine humid" conditions rather than being immersed in liquid brine.

Time Dependence of Bacterial Activity

Even if the experimental program eventually succeeds in measuring bacterial gas generation rates, a question will remain about extrapolating short-term, laboratory-derived rates to repository compliance time scales. Linear extrapolation may be appropriate for certain physical and chemical processes, but living organisms are more complex. Can bacteria continue to be active metabolically over extremely long periods in a closed system?

Because biologists ordinarily do not deal with long times and closed systems, there seems to be no literature on this subject, especially for the halophilic environment. However, literature does exist on life in extreme environments, and there are basic principles that are true for all forms of life under all conditions.

In the closed WIPP facility, organic and inorganic nutrients must be supplied from the emplaced waste and the enclosing salt. Neither of these can be regarded as a rich source of the complete range of compounds and elements required by the organisms likely to be present. Because extremely halophilic bacteria tend to have unusually complex nutritional requirements, they are not likely to thrive in an environment with a restricted range of substrates (Kushner, 1978). Nevertheless, it is probable that limited amounts of suitable nutrients are present and therefore, that the potential exists for metabolism and cell reproduction.

Unlike the culture medium in a laboratory flask, WIPP waste would not be homogeneous. Bacteria require an aqueous medium to dissolve and transport nutrients and to disperse metabolic products. Unless the repository floods, there is no continuous aqueous medium to give full access to the scattered supply of nutrients.

If a repository room does become flooded (by human intrusion, for example), biological activity will bloom for a time. The activity cannot continue indefinitely because of the bacterial generation of waste products that inhibit further metabolism and the conversion of limiting nutrients to forms that are not available to succeeding generations of bacteria.

Consider a specific example. Phosphorus typically makes up 2-3 percent of the dry weight of a microorganism. Phosphorus is indispensable to life; it is a component of the genetic material (DNA) of every living cell. Some of this phosphorus may be recycled promptly after the death and lysis of a particular cell. However, in a closed environment such as WIPP, a small remainder may be retained in a biologically inactive form for long periods. DNA samples at least 50,000 to 100,000 years old have been isolated from several sources (Williams, 1995). The phosphate deposits of Florida and Tennessee are a striking example of the way a life process can steadily lose a critical material in a "side stream."

Similar arguments can be made concerning other elements and compounds essential for the metabolic activities of bacteria in WIPP. In a closed system, it can be anticipated that at least one of these materials eventually will become limiting, thereby halting reproduction and metabolism.

"Real Waste" Tests

The DOE WIPP program plan of 1990 called for an engineering-scale program of "real" gas evolution tests to be run in parallel with laboratory studies. The plan involved emplacing a large number of drums of actual waste in one or more underground rooms, which would then be sealed and monitored. When this program encountered both practical and political objections, it was redesigned to use specially constructed metal "bins" to be filled with actual waste, placed underground, and monitored. This program also was canceled.

What remains of the real waste concept is the Source Term Test Program set up in 1994 at Los Alamos National Laboratory (LANL; see Chapter 5). As part of this program, gas evolution from selected samples of actual waste under various conditions is to be monitored at both the drum and the liter scale. As discussed in Chapter 5, no results from this program have been reported yet.

Conclusions on Gas Generation and Gas Pressure

The committee thinks that pressurization by gas generated in an undisturbed repository is not a serious concern for three reasons:

1. Laboratory experiments have recently established that hydrogen generation by anaerobic chemical corrosion of steel can occur in WIPP, but only when liquid brine and metal come into physical contact. Gas evolved from such brine-induced corrosion is limited by the contact area of brine with the steel drums. The brine-humid conditions anticipated in the repository restrict the expected gas-generating mechanisms to fewer than the set originally considered.

2. Biological gas generation will be no more than a brief transient because it will be limited by brine availability as well as by nutrient depletion.

3. Although gas sealed in waste-filled rooms can eventually approach lithostatic pressure due to creep closure of the rooms, it will represent a small total volume and a small gas pressure energy.

ROOM CLOSURE

As a result of an extensive rock mechanics research program, prediction of creep closure of repository excavations at WIPP is relatively straightforward. The time-dependent stress-strain behavior of salt is understood sufficiently well to predict closure for most design situations (Freeze et al., 1995b; Freeze and Christian-Frear, in preparation; Munson, in press). Studies to date show agreement between predicted and observed room closure rates within less than 10 percent. Further details are presented in Appendixes C and D.

Marker Bed Considerations

The main source of brine inflow to the repository rooms would be from three adjacent anhydrite layers, two above and one below the rooms. Marker bed (MB) 139, located about 1.5 m below the 4-m-high disposal room, is about 0.9 m thick. Anhydrite "a" and anhydrite "b" (each 0.1-0.2 m thick) are some 2 to 4 m above the 4-m-high disposal room, and MB 138 (approximately 0.2 m thick) is about 11 m above the ceiling (see Figures 3.1, 3.2; and Beauheim et al., 1993a).

The rooms are wide enough (10 m) that the normal rock fracturing and collapse processes around the rooms (i.e., formation of the DRZ) may cause these layers to be brought into hydrological connection with the waste rooms. Over a period of several years from the time of excavation, the DRZ will extend progressively to a distance above and below the repository comparable to the half-width of the room (i.e., about 5 m above and 5 m below the room).

The waste emplacement room dimensions (4 m high × 10 m wide × 91 m long) and the proximity of marker beds suggest that the DRZ formed around each room (including roof sag, floor uplift, and wall fractures) could, after some time, provide hydrologic contact between the waste-filled rooms and one or more marker beds.

Hydrologic contact is a matter of concern because it could create a path for fluid transfer between adjacent rooms. Even when undisturbed, the marker beds are more permeable than the neighboring halite. The

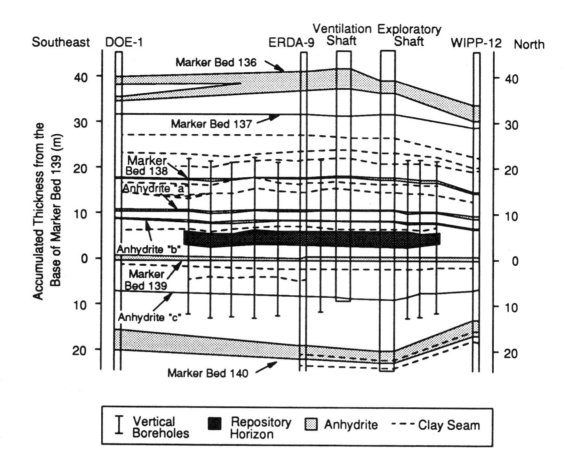

FIGURE 3.1 Stratigraphy of the Salado Formation near the repository horizon, showing adjacent anhydrite layers. Source: Freeze et al.(1995a).

"best-estimate" anhydrite permeability is of the order of 10^{-19} m^2, which is two orders of magnitude greater than the corresponding best estimate permeability for undisturbed halite (Freeze et al., 1995a, b). (See the discussion below on Backfilling and Compartmentation.)

If this interconnection should develop while the gas pressure in the WIPP excavation is below the fluid pressure in the adjacent rock, brine will flow in from the marker beds. The amount of brine available is theoretically large because the MB strata are areally extensive. However, the absolute quantity of Salado Formation brine that can enter an excavation in the repository horizon is limited by the remaining void

volume of unconsolidated material during creep closure of the salt (see Appendixes C and D).

Should brine come in contact with the steel drums, gas generation would occur. However, several effects, identified and discussed below, suggest that this gas pressure buildup would be limited.

1. A pressure rise due to gas generation will reduce the rate of flow into the excavations. If the gas pressure reaches lithostatic, it would open fractures in the marker beds, greatly increasing the marker bed permeability to outward flow from the rooms. Gas would tend to flow out of the rooms.

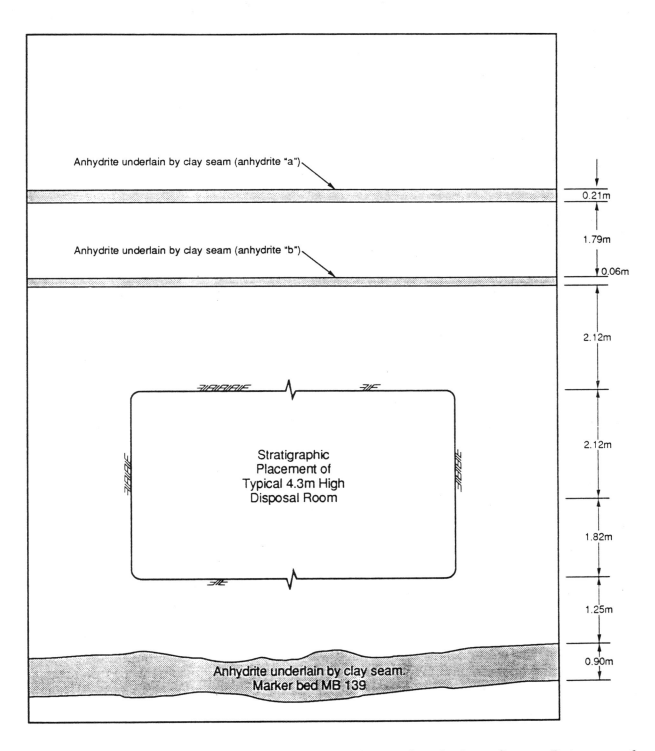

FIGURE 3.2 Close-up of stratigraphy of the Salado Formation at the repository horizon. Source: Department of Energy (1991, Figure 1-4).

Another possible effect of an intersection of the DRZ with the marker beds (particularly those above the repository) is the escape of gas by two-phase flow into the coarser fractures of the marker bed. This becomes appreciable only if the gas-brine pressure approaches lithostatic, when opening of fractures in the anhydrite layers would occur, resulting in a substantial increase in permeability of the marker beds. In this case, as shown by Webb and Larson (1996, Figure 4.9d, p. 37), the 1° inclination of the marker beds results in countercurrent flow (i.e., brine flows into the excavation as gas flows out), increasing the amount of brine inflow. Although variable, the increased inflow is of the same order of magnitude as flow into the unpressurized open excavation.

At gas pressures below lithostatic (i.e., without interbed fracturing), countercurrent flow is not significant. Furthermore, gas pressure acts to reduce brine inflow.

2. During earlier stages of room closure (i.e., 50-100 years), the floor uplift in each waste-filled room would tend to isolate the waste from liquid brine. During this period, the waste drums would be subjected to only a moderately humid environment, in which, as discussed in the previous section, gas generation by drum corrosion is considered negligibly small. However, if the brine remains in communication with the waste-filled drums, the desirability and feasibility of engineered designs to reduce or eliminate return of entrained brine to the waste rooms should be considered.

THE COMBINED EFFECTS OF BRINE INFLOW, GAS GENERATION, AND ROOM CLOSURE

The preceding discussion indicates that the predicted consequences of the combined effects of creep closure of the waste-filled disposal room, formation (and eventual healing) of the disturbed rock zone (DRZ), brine inflow, and gas generation can vary considerably depending on the particular assumptions made with respect to each of these variables (see also Freeze et al., 1995a). From the arguments outlined earlier in this chapter, the committee concludes that the amounts of brine and pressurized gas are likely to be small.

To assess the overall consequences of these issues on radionuclide isolation, it is necessary to incorporate the "room effects" into the entire network of excavations, including the shafts and seals. DOE has made the very conservative assumption that all waste-filled rooms remain permanently interconnected via a relatively high permeability DRZ (see Chapter 2) There is strong evidence, however, that placement of crushed salt seals between rooms and back filling of all other underground excavations with salt from mining operations will result in an overall low permeability within the excavation close to that of the intact salt (of the order of 10^{-20} m^2 or lower) within a few hundred years (see Callahan et al. [1996] and Appendix C). Each waste-filled room will then be an isolated compartment within the essentially impermeable Salado Formation. This compartmentation would greatly reduce the probability and consequences of an E1E2 borehole intrusion (see Chapter 2) and also the potential for brine flow up the sealed shafts.

Backfilling and Compartmentation

The development of hydrological contact between the marker beds and the waste-filled rooms could be inhibited somewhat by the use of well-placed (e.g., pneumatically injected) crushed salt backfill in the waste-filled rooms. Although this probably will not reduce the early development of the DRZ significantly because of the difficulty of packing the fill to a high enough density to obtain a rapid build-up of back pressure in the early stages of room closure, sufficient back pressure eventually will develop to reduce the aperture and hence the permeability of the marker beds.

Hydrological communication between the marker beds and the waste-filled rooms is of little significance provided hydrological communication between rooms is prevented. Careful design of entry rooms and early installation of room and panel seals to ensure that waste-filled rooms are isolated effectively from each other, or compartmented—as discussed in Chapter 4— could greatly reduce the vulnerability of a WIPP repository to human intrusion by drilling (see Chapter 2) or by water injection (see Box 3.2).

BOX 3.2 Potential Consequences of Brine Injection from Petroleum Recovery Operations on Repository Inflow

Although the permeability of halite in the Salado Formation is probably negligible, there remains some concern about flow through marker beds and other impurities in the formation. In oil-producing regions such as southeastern New Mexico, it is common to inject fluids into the deep subsurface, either to stimulate secondary recovery of oil in partly depleted oil reservoirs (e.g., by waterflooding) or to dispose of the large volumes of brine that typically are produced simultaneously with oil. Both types of operations involve high-pressure injection of fluids into a deep rock formation. If there is a failure in the well casing or in the grout or cement outside the casing, fluid can leak into overlying formations and flow laterally along one of the many anhydrite layers in the Salado.

One such recent occurrence, in the Rhodes Yates Field of Lea County in southeastern New Mexico, apparently caused significant flow through the Salado (reported as an unexpected flow of over 1,000 barrels per hour encountered by a well being drilled in 1991; see Silva, 1994, 1996). The unexpected flow through the Salado was attributed to injection into old (1940s-1950s) production wells in the Yates Formation, some 200 m below and 3 km away from the Salado horizon in the well where the flow was encountered. The flow probably was transmitted through anhydrite marker beds.

Because injection wells operate close to the WIPP site now and similar operations can reasonably be anticipated in the future, accidental leaks from such operations might represent a source of fluid that could migrate into the repository. Brine volumes under such conditions would be much greater than those that would seep into the repository from the Salado under undisturbed conditions. The committee believes that the likelihood and consequences of this type of disturbance occurring at the WIPP site should be evaluated by DOE.

There are differences between the petroleum production operations in the vicinity of WIPP and those in the Rhodes Yates Field (see Fig. A.6, Appendix A). Near WIPP, oil is produced below the 500-m-thick Castile anhydrite and salt formation, which underlies the Salado. The main gas-producing horizons are approximately 600 m deeper. Thus, injection for secondary recovery would be 700 m or more below the repository horizon. At Rhodes Yates, the Salado lies unconformably on the oil-producing Guadalupian series. The Castile Formation is absent.

Given the current improved well completion practices and the presence of the Castile into which fugitive brine may flow before reaching the Salado, the probability that fluid injected for enhanced recovery near WIPP would invade the repository horizon would seem to be lower than the Rhodes Yates incident. The potential of such a leak due to the injection of brine for disposal directly into the Castile, closer to the repository, for the period during which petroleum production at WIPP is likely to continue, is of greater concern. The committee recommends that DOE demonstrate how the performance of the repository will be assured against fluid injections.

Conclusion on the Combined Effects of Brine Inflow, Gas Generation, and Room Closure

The initial brine inflow into rooms is likely to be low enough that rates of gas generation will also be very low. Gas pressure will develop slowly (over hundreds of years) and more rapidly if the void volume of the room is reduced by backfill. Eventually, that is, in some hundreds of years, salt creep will close the rooms around the waste containers, with gas isolated in the voids between the waste in individual rooms and pressurized to the order of the lithostatic pressure. The hazard of such "pockets" of pressurized gas is considered by the committee to be negligible because of its small volume and small gas pressure energy. The committee believes that placement of effective room and panel seals will reduce greatly the potential for

hydrological communication between the individual rooms.

SUMMARY AND CONCLUSIONS

After some hundreds to perhaps a thousand years, the steady-state condition of the undisturbed repository will be one in which the residual void in each room is a small fraction (10-15 percent) of the original room volume. This residual volume will be occupied by a mixture of deoxygenated gases, mostly hydrogen, probably at a pressure approximating lithostatic. These conditions would be established by Darcy flow of brine along the interbeds, controlled by room gas pressure and interbed permeability. In the absence of adventitious brine, only "brine humid" conditions are expected. Creep closure and gas pressure buildup will serve to limit liquid brine entry.

Waste-filled rooms would probably be isolated from each other, so that the consequences of human intrusion should be considerably less severe than for the "interconnected rooms" scenario considered in the DOE PA. Chapter 4 discusses the engineering feasibility of compartmentation and other, relatively simple, engineering measures that may be taken to further improve the overall performance of WIPP as a TRU waste repository.

Chapter 4

Engineering to Improve Predicted Repository Performance

As discussed in Chapter 2, the committee judges the probability of radionuclide release from a well-sealed Waste Isolation Pilot Plant (WIPP) repository under undisturbed conditions to be very small. Concern has been raised that corrosion of the steel waste containers exposed to brine could generate gas at substantial pressure in the repository. Although the gas itself would be essentially nonradioactive, it could entrain radioactive material in the process of being released if a drill hole penetrated the repository at some time in the future.

In examining such possibilities, it is useful to consider ways in which the likelihood of potential releases could be reduced by changing the design of the repository or the form of the waste before it is placed underground. It is technically possible, for example, to incinerate the waste at the surface, thereby eliminating the possibility of gas generation underground. However, incineration also poses some risk of radionuclide exposure and would be very costly.

The Department of Energy (DOE) has conducted a number of studies (Lappin and Hunter, 1989; Marietta et al., 1989; Butcher, 1990; DOE, 1991, 1995b)[1] to examine a variety of such options to reduce risks. A short list of engineered alternatives is discussed below.

REPOSITORY DESIGN AND EXCAVATION ALTERNATIVES

Contact of drums with brine could be substantially lower than suggested in Chapter 3 because of both the probable behavior of the excavations (see Box 4.1) and

---

[1]DOE (1991) is a comprehensive study of a variety of engineered alternatives by a group of more than 50 engineers and scientists, known as the Engineered Alternatives Task Force (EATF). Results of the EATF study are referred to later in this report. DOE (1995b) is known as the Engineered Alternatives Cost /Benefit Study (EACBS).

the use of relatively simple engineering design changes, as noted below.

Natural and Engineered Sumps

Closure of the waste storage rooms, which are designed to be 4 m high × 10 m wide, will result from roof sag and floor uplift. Development of the disturbed rock zone (DRZ) and the associated floor uplift will create a "sink" for any brine flowing into the room, while simultaneously raising the waste drums above the brine, thus minimizing any immersion of the drums were liquid brine to be present. Also, because the rooms will follow the 1° inclination (i.e., a dip of 1.8 percent) of the marker beds (MBs), brine will tend to accumulate at the lower end of each storage room, creating the possibility of a drainage sump at this end of the rooms to remove the brine from contact with waste.

Design Considerations to Minimize DRZ Effects

If entrances to each room were to consist of *two* excavations, each 3 m wide, 3 m high, and about 25 m long, separated by a wider pillar, and if the center height of the excavations was located midway between the anhydrite layers, it should be possible to reduce the DRZ for each 3-m-wide excavation to no more than about 1.5 m above and 1.5 m below the roof and floor horizons, respectively (probably considerably less if the rooms are excavated and filled rapidly, within several weeks). In this way, it should be possible to avoid disturbing the anhydrite (brine-bearing) layers or connecting them with the DRZ, thus avoiding any inflow of brine into this section of the excavations. The 25-m section could be sealed with crushed salt or prefabricated compacted salt blocks, thereby isolating each room or panel from its neighbors; these seals (free

BOX 4.1 Creep Closure and the Disturbed Rock Zone

Excavations will creep closed by movement inwards from the entire volume of salt, and eventually by a lowering (subsidence) of the surface, and/or gaps between the salt and overlying more elastic formations. A crude analogy to the viscous behavior of salt is the filling of a cavity of air made in a volume of water. The air cavity will fill by flow of water towards the cavity. The water occupying the volume of the cavity will result in a (slight) lowering of the water surface level.

Measurements at WIPP show that open rooms tend to close at a constant rate of about 1 percent per year. After about a hundred to a thousand years, then, it will be difficult to identify where the original rooms were excavated, because these rooms would be filled in by flow of the entire salt mass.

The Disturbed Rock Zone (DRZ) refers to a region, usually within one diameter or less of the walls of an excavation, in which the difference between the rock stresses prior to and after excavation is sufficient to produce microcracks in the salt. In the immediate vicinity of the walls the microcracks may coalesce to form open fractures. It has been shown, both by laboratory experiment and field observations at WIPP, that the microcracks and fractures can be healed when the deformation of the salt towards the excavation is resisted by a 'back-pressure.' This back-pressure can be generated by placement of seals or by back-filling the excavations with crushed salt. Calculations (Callahan et al., 1996) indicate that the permeability of a shaft seal of crushed salt that has been dynamically compacted to 90 percent of the density of intact salt (Hansen and Ahrens, 1996; Brodsky et al., 1996) will achieve a value close to that of the intact salt within 50-100 years. The DRZ will disappear over the same time period. Back-filled rooms will require longer to come to equilibrium at lithostatic pressure, but should certainly reach a condition similar to the shaft within a few hundred years.

Some redesign of the room and pillar layout may be needed if room compartmentation is introduced. The general principle of designing operations to minimize disturbance of the marker beds prior to room sealing is a useful one to follow. Such seals also should reduce potential impacts of the E1E2 human intrusion scenario substantially (see Chapter 2). Because the projected inventory of transuranic (TRU) waste destined for WIPP may be less than the design capacity of the repository (DOE, 1995c), the separation of individual rooms could be increased and the room seal designs improved.

Room Seals, Panel Seals, and Backfill

Once filled with waste and backfilled, the rooms should close within about 100 years, given the absence of brine in the DRZ (see Appendixes C and D). Although a central "core" of crushed waste materials in the rooms will probably retain significant local permeability, these materials can be isolated from each other by effective room and panel seals. There seems to be no reason why such seals cannot be designed to be as effective as shaft seals. Thus, the room and panel seal permeabilities would decrease progressively (e.g., to 10^{-16} m^2 or lower within 100 years), as assumed for the shaft seals (see discussion below), and, given the essentially viscous long-term behavior of salt, would approach the undisturbed value for salt (i.e., essentially zero) after 500 to 1,000 years. Thus, for at least 9,000 of the 10,000 years for which E1E2 intrusions must be considered, individual waste-filled rooms would be isolated from each other.

The effectiveness of room and panel seals in reducing radionuclide releases has been noted in DOE reports:

> *The concept of sealing individual rooms or portions of rooms, using thick salt 'dikes' which isolate smaller volumes of waste from each other, was considered the most feasible facility design alternative . . . .* (DOE, 1991, p. A-39)

and

> *This alternative was considered for mitigating the effects of the two-borehole* [i.e.,

of brine flow into the DRZ) should close as rapidly and tightly as the shaft seals.

the E1E2] *scenario, and to a limited extent, the single borehole drilled into the Castile brine. The EAMP* [Engineered Alternatives Multidisciplinary Panel] *modified this alternative by suggesting that floor to ceiling salt seals could be installed at each end of the waste disposal rooms, as well as at appropriate locations within the rooms. This would decrease the effective permeability of each waste disposal panel, and prevent hydraulic communication between the two boreholes. If this alternative is implemented, it would appear to effectively eliminate the effects of the two-borehole scenario . . . .* (DOE, 1991, p. A-30)

A final report of the Engineered Alternatives Cost/ Benefit Study (EACBS) (DOE, 1995b) considers a wide variety of engineered alternatives, ranging from the use of backfill around the waste drums to reduce the void volume (and, hence, the time required to compact and seal in the waste) to the sealing of individual rooms. The prudence of considering relatively simple measures such as room seals has been recognized in concluding statements such as the following: "Communication between the rooms during an intrusion scenario is significantly reduced (gas, brine, and radionuclides)" (DOE, 1995b, p. B-7).

The committee is disappointed that, as of the time of preparation of this report, these engineered alternatives, despite their appearance in the Engineered Alternatives Task Force (EATF) and EACBS reports and their benefits being noted in Systems Prioritization Method (SPM) efforts (Sandia National Laboratories, 1995), have been neither evaluated by the WIPP program in past compliance documents nor incorporated into the 1995 performance assessment process (DOE, 1995a). The 1995 baseline model used (DOE, 1995b, Introduction) for compliance purposes is a connected repository, a suboptimal design (see Chapter 2). The benefits of using backfill are also supported by recent SPM results (Prindle et al., 1996).

Pre-Emptive Mining of Potash

The McNutt Member of the Salado Formation above the repository horizon contains potash that could be attractive for mining in the future. The field extraction methods currently used in New Mexico potash mines result in substantial ground deformation, both above and below the mined-out horizon, and surface subsidence. The ground deformation and subsidence over the mined-out area involve fracturing of the overlying strata, such as the Culebra Dolomite and the Dewey Lake Red Beds. This, in turn, may increase the hydraulic transmissivity of these layers significantly.

It is not yet clear how important this increased transmissivity might be in compromising the ability of the Culebra and associated strata to delay release of radionuclides from the repository. Studies by DOE are in progress. Should it be determined that the effects are sufficiently adverse to warrant corrective action, it is technically possible, although expensive, to extract the potash "pre-emptively" (i.e., as part of the engineering design of the repository) in such a way as to avoid subsidence and associated damage to the overlying strata.

Extraction of potash would be carried out in two stages, using a room and pillar method with cemented backfill (Brady and Brown, 1987). In the first stage, approximately half of the potash is mined across the entire minable area. The rooms are then filled with cemented backfill. In the second stage, the ore remaining between the cement-filled areas is extracted, and the voids are again filled with cemented fill. In this way, deformation of the overlying formations would be negligible. If desired, special backfill material could serve as a protective cap over the repository and as a deterrent to future drilling.

Waste Form Modification

Concern about potential gas generation in WIPP (Bredehoeft, 1988) arose from the possibility that large quantities of brine might accumulate in the repository during waste storage (see Appendix C). The 1991 EATF (DOE, 1991), a multi-disciplinary group of over fifty scientists and engineers, considered a wide range of waste form alternatives based on existing technologies or on technologies that could be developed within a few years. Such technologies include methods to shred; compact; cement or grout; incinerate; vitrify; melt metal (producing ingots or slag); add pH buffers, salt, or sorbents; decontaminate metals; and use noncorrosive containers. Changes in facility design also were considered. These options were screened to fourteen combinations that were chosen for detailed evaluation. Although such technologies either were available or probably could improve the waste form, the 1991 EATF report found that

• costs would be higher than if wastes were untreated,
• worker risk would be higher than for untreated waste handling,
• a waste treatment system would take significant time to implement, and
• treatment options would involve more complicated regulatory requirements.

In general, waste treatment options were viewed as properly held in reserve in case a need to supplement more readily implementable alternatives, such as backfill, arose. The 1995 EACBS (DOE, 1995b) chose to evaluate eighteen alternatives and reached similar conclusions and provided a specific ranking of the various alternatives.

In the committee's view, the conclusions of the EATF and EACBS are appropriate. As discussed in Chapter 3, gas generation is not anticipated to be a significant problem at WIPP; therefore, the use of advanced treatments such as incineration is unwarranted.

Findings, Conclusions, and Recommendations on Engineered Features

The committee believes that the WIPP design should incorporate engineered alternative features, and the performance assessment (PA) should reanalyze the human intrusion scenarios, incorporating (1) the existence of effective room/panel seals, and (2) a DRZ permeability that decreases with time, consistent with the known behavior of salt. Further, it is feasible, probably with no major increase in cost, to design the repository so that waste containers avoid contact with brine, thereby essentially eliminating gas generation, and to ensure that waste panels are isolated from each other effectively by seals that are comparable in permeability (10^{-16} m^2) to shaft seals.

The committee thinks that it is prudent to consider such engineering measures as backfilling of the waste rooms to ensure early reconsolidation of the DRZ and minimization of brine inflow. More advanced options, such as incineration or vitrification of TRU waste, are unwarranted.

SEALING OF SHAFTS AND BOREHOLES

In the committee's view, the flow of radionuclide-contaminated brine through inadequately sealed vertical shaft excavations at WIPP is the most probable pathway for release of radionuclides from the repository to the accessible environment under *undisturbed* conditions. Calculations (DOE, 1995d; and Appendix C) indicate that a seal permeability of 10^{-16} m^2 or less, over a minimum seal height of 100 m, is needed over the total (100 m^2) cross-sectional area of the four shafts to limit brine flow sufficiently through the seals to provide adequate isolation of the repository for 10,000 years. Figure 4.1 shows the results of PA calculations on the cumulative flow through the seals in 10,000 years.

In the figure, the net total brine flow through a 100-m-high seal of crushed salt (in each shaft) is plotted for various assumed values of the seal permeability. Each of the different flow values shown for a given seal permeability is the result of one "run" (or CCDF "realization"—see Appendix B) of the repository seal model. This model involves a number of parameters such as salt permeability, brine inflow to the rooms, gas generation, room closure, and gas pressure. Values for each parameter were selected at random from a range of values, and the model was run, computing for the particular combination of parameters selected the mean brine pressure at the bottom of the shaft over 10,000 years. In some cases, this mean pressure over 10,000 years was less than the net flow, or "cumulative release" was negative, that is, brine flowed *into* the repository. In other cases, the reverse was true, that is, the mean pressure below the shaft seal was greater than hydrostatic and so brine flowed *out* of the repository. Thus, Figure 4.1 shows both positive (outflow) and negative (inflow) values at each permeability tested. For permeability of 10^{-16} m^2 or lower, the flows were negligibly small in all cases tested.

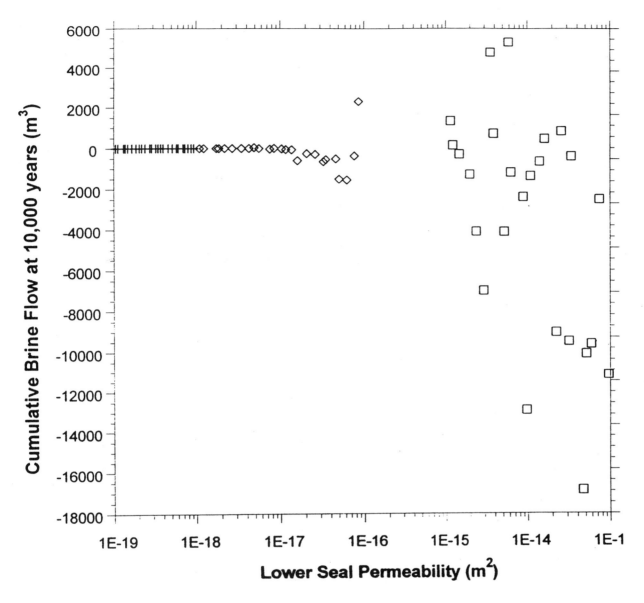

FIGURE 4.1 Predicted cumulative brine flow through (100-m-high) lower seals of the WIPP shaft. Source: Department of Energy (1995d, Figure C-1, p. C-5). This is a compilation of three sets of PA sampling calculations varying the Lower Seal Permeability. Squares indicate the cumulative flow for seal permeabilities in the range 10^{-13} to 10^{-15} m^2; diamonds, in the range 10^{-16} to 10^{-18} m^2; and plus signs, 10^{-18} to 10^{-19} m^2. In the calculation, the rate of brine flow through the seal depends on the *difference* between the pressure below and above the seal and lessens as the seal permeability decreases. The negative results in the figure indicate that the average pressure generated in the repository over the first 10,000 years is less than the pressure (hydrostatic) above the seal—so fluid flows *into* the repository. Thus, the Cumulative Brine (out) Flow is negative (i.e., it is a net inflow). When the permeability is 10^{-16} m^2 or less, there is little flow in or out. These PA calculations are based on the assumption that there are no designed seals in any of the underground excavations. Thus, shaft seals alone are sufficient to prevent radionuclide release for the undisturbed case.

It is seen that a lower seal permeability (abscissa) of 10^{-16} m^2 or less is needed to reduce the cumulative brine flow at 10,000 years (ordinate) to negligible levels for all repository parameter combinations. The accuracy of this result is shown by independent calculations performed by the committee and given in Appendix C. Similarly, a permeability of 10^{-18} m^2 or less is needed to limit gas flow through the seal (DOE, 1995d).[2]

The four shafts at WIPP are each approximately 6.1 m (20 ft) in diameter and approximately 660 m (2,160 ft) deep. The upper 256 m (840 ft) of the shafts passes through relatively permeable water-bearing sediments consisting of evaporites, carbonates, and clastic rocks. The remaining 404 m (1,325 ft) to the bottom of the shaft (and the repository horizon) consists of bedded salt in the Salado Formation. The Salado also extends approximately 200 m below the repository horizon.

Once the repository has been filled with TRU waste, the entire 660-m column of each shaft will be sealed with a variety of seal materials, as shown in Figure 4.2. In designing the seal system, it is necessary to incorporate features that effectively will prevent vertical flow of fluid both within the shaft itself and within an annular region extending about one-half to one shaft radius from the shaft wall into the rock. Within this region, referred to as the DRZ, the initially intact, essentially impermeable salt has been fractured as a consequence of rock stress changes introduced by the process of excavation and by the continuing changes that have occurred over the decade or so since the WIPP shafts were excavated (Figure 4.3). Given the additional 30-40 years needed to fill the repository, these time-dependent effects would be operative for a total of approximately 50 years before shaft sealing could start, creating an annulus of relatively high permeability around the shaft (Figure 4.3). This annular pathway for fluid flow must be eliminated to seal the shafts effectively.

---

[2]Although concentrations of hazardous gases are not likely to be sufficient to exceed Resource Conservation and Recovery Act regulatory limits on gas leakage to the accessible environment, the gas pressure of the repository is an important design consideration. Any pressurized gas in fractures in the DRZ around the shaft and in the interstices of the crushed salt used to seal the shaft could reduce the rate of salt recompaction significantly, both in the seal and in the DRZ. Brine within the same interstices could have a similarly retarding effect on salt consolidation.

Development and Healing of the Disturbed Rock Zone

The "damage" induced in the DRZ varies progressively from small, isolated microcracks in the regions of low stress change remote from the shaft, through a coalescence of the microcracks, to extensive through-going vertical fractures in the immediate vicinity of the shaft wall. Only in this more extensively fractured region is shaft permeability increased significantly.

Recent gas and brine flow permeability tests (Dale and Hurtado, 1996) around the Air Intake Shaft at the 660-m level and at the top of the Salado (256 m) reveal that there is no measurable change in permeability of the salt at either location, except within an annulus extending approximately 1 m from the shaft wall.[3] Thus, the region of increased permeability is likely to be a small part (10 to 20 percent [maximum] of the shaft radius) of the DRZ around each shaft.

Laboratory studies (Brodsky, 1995) clearly indicate that stress-induced fractures in salt specimens can be eliminated when the damaged region is subjected to sustained moderate confining pressure (approximately 5 MPa [MegaPascals]). The permeability of the cracked specimens is found to approach that of intact salt within a few days.

In the full-scale shaft sealing situation, the rate of reclosing or "healing" of induced (permeable) fracture zones around a shaft in the Salado will depend on the rate of buildup of the confining stress. This depends, in turn, on the compressibility of the shaft seal material and the rate of steady-state creep of salt around the shaft (Figure 4.4).

Numerical simulations by DOE contractors indicate that the width of the DRZ annulus at the time the shaft is filled (i.e., 50 years after excavation) will vary from $0.6R$ (where R [= 6.1 m] is the shaft radius) at a depth of 256 m (i.e., at the top of the Salado, where the lithostatic pressure is 6 MPa) to approximately $0.85R$ at 660-m depth (i.e., at the bottom of the shaft, where

---

[3]Because the rate of increase in the extent of the DRZ decreases exponentially with time, the 1 m of DRZ developed in about 10 years (the approximate age of the Air Intake Shaft) would grow to no more than about 2 m in 50 years.

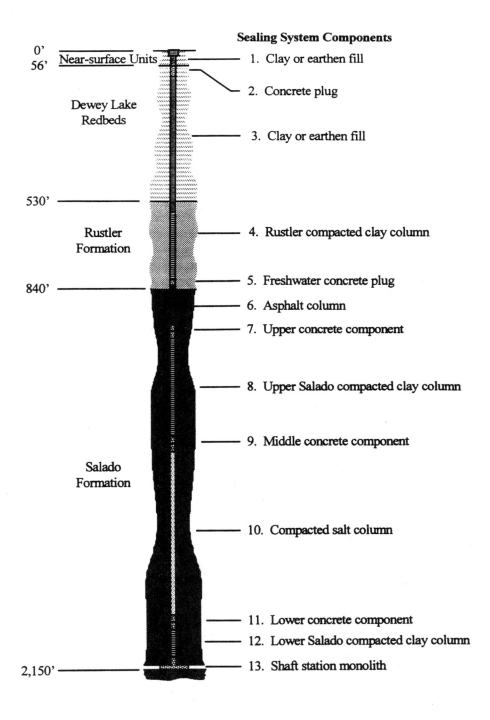

FIGURE 4.2 Arrangement of the WIPP shaft sealing system. Source: Department of Energy (1995d, Figure 2-1, p. 19).

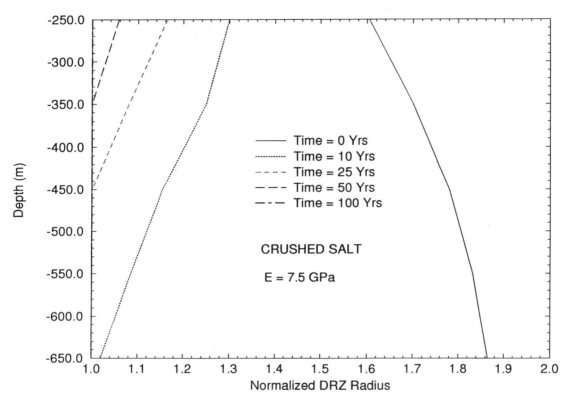

FIGURE 4.3 Calculated radial extent of DRZ annulus around a seal of crushed salt, compacted to 90 percent of the intact salt density. Source: RE/SPEC calculations done under contract to the Sandia WIPP Shaft Sealing Program, in RSI Calculation File 325/11/05.

the lithostatic pressure is 15 MPa). Computations (see Figures 4.3 and 4.4) indicate that with crushed salt that is compacted to 90 percent of the density of intact salt, the DRZ annulus will, 50 years after filling the shaft, have been reduced to $0.05R$ at the top of the Salado and totally eliminated between 350 m and the bottom of the shaft (660 m), that is, over almost 300 m. After 50 more years, the DRZ is eliminated over the entire 400-m Salado section. Because the region of increased permeability in the DRZ is confined to the portion immediately around the shaft wall, it seems probable that the entire 400-m Salado section in the shaft will achieve a permeability close to that of intact salt within 50 years of filling the shaft.

The computations above do not allow for the delaying effect of adventitious brine or high-pressure gases that could be introduced into the DRZ fractures.

As mentioned earlier, the presence of these fluids would inhibit closure of both the DRZ and the compacted salt. Eventually, the connected, fluid-filled fractures in the DRZ (and regions in the crushed salt) would probably degenerate into isolated spherical inclusions containing gas or brine at lithostatic pressure within the "healed" salt, but the elimination of permeability may be delayed substantially. It is therefore desirable to ensure that the DRZ fractures and intergranular regions of compacted salt are maintained free of brine or gas during the 50-100 years required for full compaction under dry conditions.

To this end, several strategies have been proposed, including the introduction of supplementary seals above and below the region of the compacted salt seal. Use of the concrete seals, which have a significantly higher initial stiffness than the compacted salt, will result in a

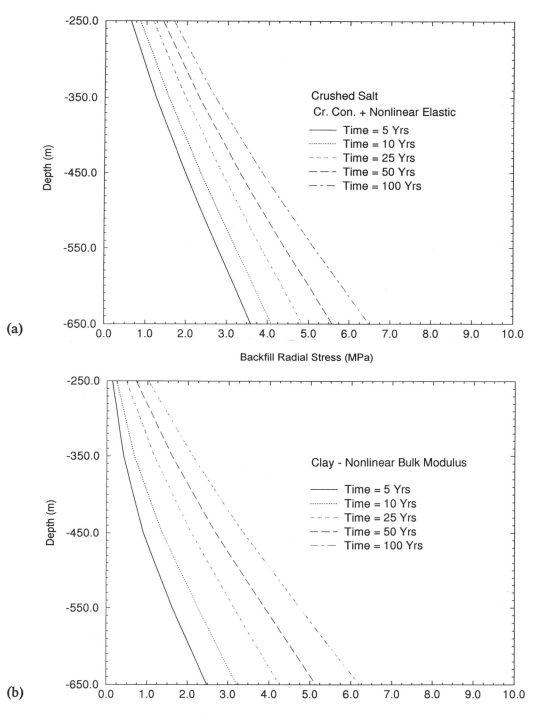

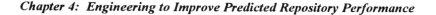

FIGURE 4.4 Radial stresses (back-pressure) in (a) compacted crushed salt backfill, and (b) compacted clay in the WIPP shafts as a function of time and depth. Source: RE/SPEC calculations done under contract to the Sandia WIPP Shaft Sealing Program, in RSI Calculation Files 325/11/03, 325/11/04, and in RE/SPEC External Memorandum RSI(RCO)-325/11/95/23 (November 1995).

more rapid buildup of back-pressure and correspondingly more rapid elimination of any permeability in the DRZ above and below the compacted salt section, thereby sealing this section early against brine inflow. Brine and gas inflow to the compacted salt column can be eliminated effectively by placing a compacted clay column above and below the compacted salt (Figure 4.2). Although probably unnecessary, it is feasible, at the time of sealing, to "ream out" this inner portion (i.e., about 10-20 percent of the shaft radius, as noted earlier) of the DRZ above and below the compacted salt sections where the concrete is to be placed. The reamed-out sections then would be filled with rapid-curing concrete as soon as possible after reaming, ensuring very early protection of the DRZ in the compacted salt section against inflow of brine.

The benefit of retarding release of radionuclides by placing compacted clay columns at various sections in the shaft (Figure 4.2) should be considered fully in the design and PA analysis of the effectiveness of shaft seals. Present PA sensitivity studies model shaft seal permeabilities but not extensive, detailed shaft designs.

No large-scale demonstrations of the effectiveness of proposed shaft-sealing designs have been carried out. However, the reality of salt creep is indisputable; salt around the shaft definitely will flow to seal in the shaft plug and reheal the disturbed rock zone. Because DOE proposes to use compacted crushed salt as the shaft seal material, this too should reconsolidate eventually to a density and impermeability comparable to that of intact salt. The practical feasibility of using temporary seals to eliminate water from the permanent seal sections during closure has to be examined and demonstrated carefully. Because these seals will be required only at the end of the operational phase of WIPP, there is adequate time to address this design problem.

In the proposed multicomponent shaft seal system (illustrated in Figure 4.2), reconsolidated salt is expected to constitute the major long-term seal component. Ahrens and Hansen (1995) demonstrated the feasibility of large-scale dynamic compaction of crushed natural salt in a shaft configuration to achieve 90 percent of its intact density, providing greater confidence in the efficiency of its reconsolidation to produce an effective long-term seal.

As noted previously, major short-term concrete seal components are expected to provide structural support as well as to retard brine or gas flow substantially during the intervening period before the salt seal components are adequately consolidated. This performance of the concrete component has been documented recently by the Department of Energy (1995d) and, in more detail, in individual reports (Stormont, 1984, 1986, 1988; Van Sambeek et al., 1993). The importance of designing concrete to be chemically compatible and durable in the salt-dominated WIPP horizon has been emphasized by several investigators (e.g., Roy et al., 1983, 1985; Wakeley et al., 1993, 1994, 1995). In-place permeabilities of experimental salt-saturated concrete plugs (1-m diameter) in WIPP showed decreased permeabilities with time (less than 4×10^{-19} m^2) after six years (Wakeley et al., 1993). Some remaining concerns exist regarding compatibility of concrete with magnesium-rich brines found at WIPP. Additional confidence in the performance of concrete is provided by field studies of sealing methods in salt or potash mines (Eyermann et al., 1995). The stoppage of major (10 m^3/min) flow of pressurized brine in the Rocanville Potash Mine, Saskatchewan, is a good example of the ability to seal excavations effectively in potash (or salt) using current technology.

The seal design incorporates additional components of shaft seals, planned to perform individual functions (DOE, 1995d). These seals include compacted clays (Meyer and Howard, 1983) for use in both non-Salado and Salado members, asphalt to act as a water-stop between concrete members, and possibly other impermeable chemical seal components to act essentially as an O-ring. Issues of concern about the potential use of asphalt as a seal component have not been resolved fully. As with many organic-based materials, there are issues of potential degradation and gas generation, including microbial attack. Addition of lime to the asphalt mix has been suggested as a possible remedial measure (DOE, 1995d). The influence of the different compressibilities of the seal material (here crushed salt and clay) on the rate of buildup of radial stresses in the seal and the shaft wall is illustrated in Figure 4.4.

The sealing of exploratory and other boreholes, as necessary, is expected to draw on extensive experience with cementing and plugging technology in the oil and gas industry, much of which has involved bedded salt environments. Impressive early tests of the performance of the borehole plug (cementitious materials) under high well pressure conditions in the specific WIPP environment (Gulick et al., 1980) have been followed by other studies, demonstrations, and plug material development (Finley and Tillerson, 1992; Wakeley et al., 1995).

SEALING OF ROOMS AND PANELS

Sealing sections of horizontal drifts presents essentially the same problems as sealing vertical shafts, except that in this case, gravity acts to accentuate development of the DRZ in the roof and to retard it in the floor. One difference is that during placement, it is difficult to compact crushed salt in a horizontal drift to the same density as in a vertical shaft.

Salt creep should compact horizontal seals eventually to the same high level of impermeability (for both the DRZ and the excavation) as in the shafts. If it is assumed, for example, that 100 m of drift can be packed with crushed salt at 70 percent of the intact salt density (i.e., rather than the 90 percent density for the shafts), 200 or 300 years probably will be required to reach 10^{-16} m^2 seal permeability, rather than the 50-100 years needed for the shaft. Technologies of placing drift seals to isolate radioactive waste have been, or are being, examined elsewhere, for example, in Stripa (Sweden) and Pinawa (Atomic Energy of Canada, Ltd., Canada). Salt has the distinct advantage over most other rocks of exhibiting creep, allowing rehealing of the DRZ, and progressive reduction of seal permeability.

Thus, it seems technically feasible to provide, between the waste storage rooms and panels, drift seals that are sufficiently tight, perhaps 300 years after waste emplacement, to eliminate brine communication between the waste-filled rooms and panels at WIPP. Note that any such communication between rooms via anhydrite marker beds (e.g., MB 138, MB 139) would not be affected by the drift seals. These beds have a low permeability, of median value 2.5×10^{-19} m^2 (DOE, 1995a, p. 6.75) and therefore would contribute negligible brine, with regions near the excavations becoming drained during the period that the rooms are open before waste emplacement. As discussed at the beginning of this chapter, with careful room and panel seal designs, it should be possible to inhibit room-room hydrologic communication.

Whether or not drift seals contribute to the overall isolation capability of the repository depends on the credence given to the human intrusion scenarios that must be examined according to 40 CFR 194. For the undisturbed case, PA studies by DOE in which no drift seals are assumed show that the shaft seals alone are sufficient to prevent radionuclide release. Thus, for the undisturbed case, drift seals are not necessary to demonstrate compliance (see discussion of Figure 4.1).

Currently it is assumed, for example, as part of the E1E2 scenario (see Chapter 2), that brine from a pressurized reservoir will flow through the repository, carrying radionuclide-contaminated brine to the surface via a second borehole. Provision of drift seals would reduce the possibility of establishing such a flow path and hence would considerably reduce the significance of this "two-hole" type of human intrusion.

CONCLUSIONS

The ability of salt to flow under stress is a particularly valuable attribute with respect to sealing excavations in the WIPP repository. While careful attention to good practice during seal construction is very important, it appears to the committee that the design of effective shaft and panel or room seals at WIPP is entirely possible by using currently available technology. The shaft seal design system proposed for WIPP (DOE, 1995d) includes large factors of conservatism and redundancy, which ensure that the requirements of 40 CFR 191 with respect to sealing of excavations at WIPP can be met by a large margin.[4] Although room and panel seals are not essential for adequate waste isolation at WIPP given undisturbed conditions, they would reduce the significance of an E1E2 intrusion event considerably.

---

[4]This is a properly conservative design of the engineered seal. There is no contradiction between the conservatism advocated here and the critical remarks against conservatism in Chapter 2. The latter were levied against unrealistic assumptions of salt performance that could lead to a mischaracterization of how WIPP could perform.

Chapter 5

Actinide Source Term

In the performance assessment (PA) of the repository described in Chapter 2, compliance with the Environmental Protection Agency (EPA) requirements is discussed relative to two scenarios, E1 and E1E2 (Figures 2.4 and 2.5), in which pressurized brine from below the Salado Formation flows into the repository and then either directly to the surface or through the Culebra Dolomite to the accessible environment. The amount of radioactive material that might move into the environment obviously will depend on the amounts of the five actinide elements (plutonium [Pu], americium [Am], neptunium [Np], uranium [U], and thorium [Th]) that the flowing brine can transport. This quantity is the actinide source term (AST), defined more specifically by scientists at Sandia National Laboratories as the cumulative amounts, concentrations in solution, and chemical nature of the radioactive materials that conceivably could be moved in solution or suspension from the Waste Isolation Pilot Plant (WIPP) repository into the environment.

Clearly, an estimate of the AST is an important part of any performance assessment. Although the AST program includes an evaluation of five actinides, the most important issue for periods greater than a thousand years is the fate of the more abundant, long-lived isotopes of plutonium (e.g., Pu-239 and Pu-240; see Table 1.1). The study of the other actinides is important because they provide a basis for the oxidation state model in which plutonium may assume a variety of oxidation states. As an example, the following elements may be used to model the behavior of plutonium: $Am(III)$[1] for $Pu(III)$, $Th(IV)$ for $Pu(IV)$, $Np(V)O_2^+$ for $Pu(V)$, and $U(VI)O_2^{2+}$ for $Pu(VI)$ (Novak, 1995b).

Evaluation of the AST depends principally on the following:

- the total inventory of radioactive material (Table 1.1) and its decrease due to radioactive decay as a function of time;
- conditions within panels of the repository and along the route of transport (e.g., the volume of brine exposed to the waste and transported out of the repository); and
- the timing of potential release events.

The challenge in making a reasonable estimate of the AST is to gain an understanding of the coupling between the chemistry of the system and the boundary conditions and assumptions (e.g., redox potential, flow rate, gas generation, distribution of sorbing clays, presence of ligands) used in the performance assessment.

As noted in Chapter 2, performance assessments to date (e.g., that for 1992) confirm that if the repository, after completion, is left undisturbed by human activity, the presence of brine in contact with waste will not lead to migration and release of appreciable radioactivity beyond the WIPP boundary. The geologic barriers, combined with the planned engineered barriers discussed in Chapter 4, will be sufficient to prevent radionuclide escape. *Thus, for the undisturbed case, the magnitude of the AST is of no consequence for assessing compliance.*

If the repository is disturbed at some time in the future, however, it is possible that some radioactive material might escape. Therefore, estimates of possible amounts of released radioactivity are required for detailed assessment of the performance of the repository under various assumed conditions. In any such assessment, the magnitude of the AST is an important parameter.

In making a performance assessment of the scenarios described in Chapter 2, Department of Energy (DOE) scientists have undertaken an elaborate program of experimental work (Novak, 1995a) to ascertain the

---

[1] The Roman numerals specify the actinide oxidation state.

possible range of actinide concentrations in the brine as it moves from the panels in the repository to the WIPP boundary. This chapter reviews DOE's proposed experimental program (Novak, 1995b; Papenguth and Behl, 1996a, b) on the AST.

Although the committee has been given a good overview of the work on the AST that has been completed or is still in progress (see Appendix E), much of the research on the source term is not yet complete, and some important results of the program will not be available for review in the immediate future (e.g., experimental studies of Pu speciation in brine). Consequently, this chapter should be viewed as a review of work in progress and cannot serve as an evaluation of the quality or applicability of the final experimental results.

ACTINIDE SOURCE TERM MODEL

Brine that has been in contact with transuranic (TRU) waste (described in Chapter 1) likely will contain actinides as compounds or ions in solution and as colloidal particles. As the brine moves through rock units, the amount of actinide transported will change. Actinide concentrations in brine may be expected to decrease owing to precipitation of actinide-bearing phases or sorption on mineral surfaces; alternatively, actinide concentrations could increase in response to the formation of complexes in solution. To determine the amounts of actinides that might reach the compliance boundary, estimates of the following are required for each element of concern (particularly plutonium):

1. the concentration in solution, as controlled by the solubilities of the actinide-bearing phases and the kinetics of the dissolution process;
2. the amount that could form and be transported as colloidal particles; and
3. the amount that could be removed from solution by sorption on the surfaces of dolomite or other minerals (e.g., the corrensite-lined fractures of the Culebra Dolomite) as the brine moves from the repository into the surrounding rock.

DOE analyses (Sandia National Laboratories, 1992, 1995; DOE, 1995a) consistently have identified these three quantities as important to the long-term performance of WIPP. The same three quantities also have been identified as important by EPA and the Environmental Evaluation Group (EEG; Neill et al., 1996).

Estimates of these quantities can be approached theoretically with models using tabulated chemical data and basic chemical principles, and empirically through either carefully controlled laboratory experiments or measurements under actual field conditions. Each approach has advantages, and all have weaknesses. Although well-controlled experiments using ideal or model systems may provide a fundamental understanding of the chemical processes, conditions of the experiments may be simplified to the point that their direct applicability to the larger-scale, more complex conditions in a waste panel is not clear. Field-scale experiments (e.g., sorbing tracer tests) or experiments with actual waste (e.g., Source Term Test Program [STTP] experiments at the Los Alamos National Laboratory [LANL]) may be difficult to interpret, but they do provide qualitative information on potential chemical interactions and bounding estimates of actinide concentrations in brine. Confidence in the models used in performance assessment will rest inevitably on consistency in the results and the general applicability of the models to all results—experimental, theoretical, and field.

Three principal factors currently being investigated by DOE that determine actinide transport in brine—solubility, colloids, and retardation by sorption—are examined below. Only limited attention is being paid to other factors (e.g., kinetics and redox effects on the reactions) that may also finally influence the transport of actinides to the accessible environment.

Solubility

Estimating actinide solubilities from published data is particularly difficult because the electrolyte concentrations in brines are very high, whereas most existing tabulated data apply to dilute solutions. This means that the thermodynamic formalism related to solubility must incorporate the use of correction factors. Of the numerous such factors that have been proposed, the generally accepted approach has proven to be the parameters developed by Pitzer (Pitzer, 1991). These are the correction factors used by DOE scientists (see Appendix E for a more detailed discussion of Pitzer parameters).

Difficulties in determining actinide solubilities are compounded by the many gaps in published tables of thermodynamic data. DOE scientists are responsible for filling such gaps through their experimental work. A further complication is the fact that the solubilities of a number of actinide compounds are influenced by the formation of complexes, either with inorganic ions such as Cl^- and CO_3^{2-} or with C-H-O compounds from the organic materials present in much TRU waste. Add the difficulty of knowing or determining the relative proportions of the oxidation states of an actinide element in a given solution, and the complicated nature of the task that DOE researchers have set for themselves becomes apparent.[2]

Based on a review of the AST program, the committee has identified areas of concern that are described in Appendix E. Although the experimental approach chosen to estimate actinide solubilities is acceptable, it will be defensible only when supported by extensive experimental data, which were unavailable at the time of preparation of this report. When compared to published experimental data, current modeling efforts are not capable of conservatively estimating actinide solubilities. For example, Novak and Roberts (1995) found that their model for Np(V) predicted solubilities as much as an order of magnitude less than those determined experimentally. Thus, completing the experimental program in order to generate Pitzer parameters for all oxidation states of the actinides is

essential for obtaining credible estimates of dissolved actinide concentrations.

Additionally, the estimate of actinide concentrations in a brine will depend on the solubilities of actinide-bearing phases. In most of the experimental programs, too little attention is being paid to identification of the solubility-limiting phases. These phases may be unusual in composition, and thus difficult to identify, because of the complex composition of the brine and the waste inventory. Many of the thermodynamic data required by geochemical codes to model actinide concentrations in solution are either absent or contradictory. A key factor in the final result will be the "fairly subjective" choice of the solubility-limiting phase (McKinley and Savage, 1993).

Colloids

In the repository environment, many processes can lead to the formation of colloidal suspensions (Laaksoharju et al., 1995), some of which may include actinide elements. In fact, the transport of actinides by colloids greatly can exceed that anticipated from models based solely on the solubility of actinide-bearing phases (Kim, 1994). Recent results (Reed et al., 1994) have demonstrated the important potential role of colloids in retaining plutonium in suspension, with subsequent transport potentially being greater than expected (Ibaraki and Sudicky, 1995).

When a solid actinide compound is precipitated from solution, it may form suspended colloidal particles rather than a crystalline solid, or actinide ions may attach themselves to colloidal particles formed in other ways, either as inorganic particles or as complex compounds derived from organic constituents of the waste. Some of the varieties of bacteria that grow in the waste solution may themselves be particles of colloidal size and may contain sorbed actinides.

Colloid particles (usually in the size range of 10^{-9} to 10^{-5} m) generally have an electric charge, either positive or negative, that helps keep the particles apart and prevents their flocculation as a precipitate. In concentrated electrolyte solutions such as WIPP brines, the charge on particles is neutralized readily and the colloids are generally unstable. This is especially true for colloids of inorganic material (either colloids consisting entirely of actinide molecules or inorganic

---

[2] Because alpha radiation exists as a continuous source of electron acceptors (reductants), "autooxidation" by alpha bombardment is possible. In some environments this has affected redox-sensitive reactions, because Eh of the fluids is important in solubility determinations. However, the pH and Eh conditions at WIPP are controlled by the geologic environment and by constituents of the waste, particularly the metal of the waste containers and the abundant organic material in the waste. A slight change caused by alpha bombardment is expected to be negligible for both the disturbed and the undisturbed repository. Alpha-radiolysis causing oxidative corrosion is a minor concern in spent fuel storage (Shoesmith et al., 1996). Alpha-radiolysis is expected to be an even more minor concern at WIPP, because a comparison of the U.S. spent fuel inventory to the TRU waste inventory, in an isotope-specific curie comparison of transuranics only, shows spent fuel to be orders of magnitude more radioactive.

compounds with sorbed actinides) and less so for colloids in which the particles consist largely of organic material. The former are called "hard-sphere" or hydrophobic colloids; the latter, "soft-sphere" or hydrophilic colloids. In general, hard-sphere colloids readily flocculate in a strong electrolyte and, hence, are ineffective in mobilizing actinides in the WIPP environment. In contrast, some soft-sphere colloids (including colloidal-size bacteria) remain stable even in concentrated solutions and so can serve as effective transporters of actinides. It is emphasized that colloid stability is a complex function of many other variables, such as acidity and alkalinity, the nature of sorbed ions, and the kinds of organic matter and electrolytes present.

Bacteria may be the most effective vehicle for transport of actinides in the Culebra Dolomite. Their total numbers, sizes, and morphologies in indigenous Culebra brines will be determined, as well as the increase in their numbers if nutrients are added from organic material in the waste. Bacteria can influence the behavior of actinides in several ways:

1. Their metabolic activity will tend to keep conditions slightly acidic, thus helping to determine the oxidation state of actinide elements and so affecting their solubilities.

2. By sorbing ions from solution, bacteria obviously can aid in the transport of actinides.

3. The well-known tendency of microorganisms to adhere to surfaces (Marshal, 1976) may serve to retard actinide transport.

4. If significant quantities of gas are produced as a result of microbial metabolism, this gas also may influence the movement of actinides.

In addition to bacterial activity, any corrosion process will influence the pH and/or the gas content of the brine and thereby influence solubilities and subsequent transport of actinides.

The possibility of substantial transport of actinides in colloidal form has been recognized only fairly recently (Kim, 1994) and has spurred much effort by DOE scientists to investigate the complexities of colloid behavior in strong electrolyte solutions such as WIPP brines (Papenguth and Behl, 1996a). From the proposed experimental program, DOE scientists expect to learn whether there is a serious threat of active transport of actinides by colloidal particles and what measures might be taken to modify conditions in the WIPP repository to reduce this threat.

The DOE study began with a comprehensive review of recent literature on the formation and stability of colloids in electrolyte solutions and now includes detailed plans for experiments designed to answer the many remaining questions about this mode of actinide transport (see Appendix E). This effort is commended by the committee, with the expectation that these results will provide much of the new information needed to gain a deeper understanding of colloids as part of the actinide source term. The experimental work has started and is scheduled for completion soon (see Table E.1), but no results have been presented to the committee.

Retardation

If, as postulated, the dissolved and colloidal actinides that make up most of the source term move with the brine into the accessible environment through the Culebra Dolomite, their transport will be retarded to some extent by sorption on the surfaces of the dolomite and minor amounts of clay minerals (e.g., corrensite) contained in this member. Thus, the amount of radioactive material that might actually reach the accessible environment depends on the effectiveness of this sorption. Accordingly, the amount of sorption is a critical factor in any attempt to estimate whether or not the WIPP project complies with EPA regulations.

As with solubility, the study of retardation by sorption is made difficult by the high concentration of dissolved material in the brine. For dilute solutions, determining the amount of material sorbed is straightforward and experimentally simple. It is commonly expressed in terms of K_ds, or distribution coefficients, which are merely the ratios of amounts sorbed to amounts left in solution. Complexities arise with concentrated solutions, however, because (1) many substances are competing for places on any given sorption site, and (2) the nature of the sorbing substance is influenced by its speciation and the chemical conditions (e.g., acidity and alkalinity) of the solution (see discussion in Appendix E). Additionally, batch K_ds measured in the laboratory rarely have been used successfully to predict sorption on a larger field scale.

Because simple K_ds are of little use as predictors, amounts of sorption are best determined by direct experiments in which samples of natural or synthetic brine are allowed to flow through samples of the intended sorbent material, and actinide concentrations are measured before and after the passage (see Appendix E). Such experiments have been planned carefully and should give a good indication of the effectiveness of sorption in controlling actinide concentrations; however, no results of the experiments completed to date have been presented to the committee. The experimental results will have to be confirmed by results from actual field experiments. Throughout the history of the WIPP project, such studies ("sorbing tracer tests") have been planned, delayed, and canceled. The status of such tests at present is unclear. It will be difficult for DOE to take credit for sorption in the Culebra (or any other rock unit) in the absence of well-planned and completed field tests.

SCHEDULE

As noted above, the ambitious schedule DOE has set for completion of experimental work for the AST (1996) is unrealistic in that final, definitive values for some of the numbers being sought cannot possibly be obtained and evaluated critically in so short a time. Enough data may be generated to provide a reasonable basis for the proposed compliance application, but important details about the source term and its appropriate use in performance assessment will almost certainly require additional work. Strictly accurate final values are not essential for showing compliance with EPA regulations, but are nevertheless numbers for which DOE will find many uses later. The committee recommends that selected research programs be extended into the operational phase of the repository.

SUMMARY AND DISCUSSION

The actinide source term can be envisioned as the entire mass of actinides destined for disposal at WIPP, partly dissolved and partly in colloidal suspension in a brine similar in composition to Salado Formation or Castile Formation brines (or a mixture of the two). That the entire mass would ever be dissolved or suspended is hardly likely, but a minor fraction could plausibly be mobilized at some time or times in the future. Determining how large this fraction might be requires knowledge of the solubility of each of the actinide elements in brine and its behavior as a colloid, plus data on the sorption of ions and colloidal particles onto mineral surfaces with which the brine may come in contact. Data of this sort, both from the literature and from experimental work, have been and are being accumulated in impressive amounts by DOE scientists and contractors.

Data on the source term, as noted previously, will have no importance if the repository remains undisturbed by human activity for 10,000 years because no appreciable radioactivity will likely escape to the surface. However, for any of the scenarios considered in Chapter 2 describing the results of possible human disturbance of the repository, knowledge about the source term is essential for estimating possible radioactive releases from various assessments of repository performance and, hence, for making a decision on the application for a certificate of compliance to open the WIPP repository.

The ongoing experiments in DOE laboratories and DOE-sponsored universities, if carried to completion, would begin to provide the necessary information. The present schedule will not allow all the work to be finished and evaluated before the Fall of 1996. Nonetheless, enough preliminary data will be obtained to make possible a reasoned judgment of whether WIPP meets the compliance criteria for the disturbed case.

Overall, the scientific program outlined by DOE for study of the source term is adequate, provided that the program is carried to completion. Because the program at this time consists largely of work planned or in progress, it has not been possible to critically review experimental results or to judge whether these results are used appropriately in the PA analysis. Much of the AST work has not been subjected to the type of peer review that is part of publication in scientific journals, and the committee recommends that such review be sought as the work progresses. Interaction between principal investigators in the AST program and modelers in performance assessment has been disappointingly limited, and the committee recommends that this situation be corrected.

In the opinion of the committee, as soon as substantial results are available from the experiments—provided, of course, that the results show the expected low solubilities, which will permit calculation of low actinide releases based on some of the less improbable variants of the borehole scenario—then DOE will have the information on actinide solubilities needed to prepare the certification of compliance.

Experiments should continue for as long as is needed to obtain reliable values for actinide solubilities, colloidal transport, and nuclide retardation by sorption, a time during which the repository will be in full operation. The new data obtained will be important at the WIPP site as a basis for changes in operational procedures (e.g., use of backfill, improved seals, or reduced waste loading) and, more generally, for providing the fundamental understanding needed to predict actinide behavior in other situations. Actinide chemistry is important, for example, in the disposal of other kinds of radioactive waste and in the nationwide environmental remediation program that DOE has undertaken. Because the experimental work will provide basic data needed for a variety of purposes, the committee recommends that DOE continue this work beyond the scheduled completion dates in mid-1996.

There appears to be no reason that the further required work on the actinide source term cannot be accomplished simultaneously with the disposal of waste in the repository. This additional work will improve the scientific basis of the performance assessment and increase public confidence in the ability of WIPP to isolate TRU waste.

CONCLUSIONS AND RECOMMENDATIONS

The committee offers the following conclusions regarding the actinide source term:

• *For the undisturbed case, the magnitude of the actinide source term is of no consequence to the demonstration that WIPP can isolate waste from the biosphere effectively for many thousands of years.*

• *In the disturbed case, if it is necessary to rely on estimates of actinide concentrations in the brine, colloidal transport, or nuclide retardation by* *sorption, a well-documented, reviewed, scientific and technical basis for these estimates does not yet exist. With this in mind, the following observations and recommendations are made.*

With the exception of specific scientific issues raised in this chapter (and detailed in Appendix E), the scientific program outlined by DOE is adequate if the proposed program is carried to completion.

Plutonium is the principal radionuclide of interest in any release scenario. Although some aspects of its geochemical behavior can be estimated by studies of other actinides using the oxidation state model, ideally, data for each oxidation state should be followed by experiments with plutonium.

Because the AST program at this time consists mainly of proposed work, the committee has not had the opportunity to review experimental results or to evaluate whether these results were used appropriately in the performance assessment analysis. At present, much of the AST work in support of the WIPP project has not been subjected to the type of peer review that is part of publication in scientific journals. This review process is the essence of quality assurance, and the committee recommends that scientific results of the AST program continue to be published in peer-reviewed journals.

The existing schedule (see Table E.1) will not allow for the timely (i.e., by the Fall of 1996) completion of the proposed experimental work, proper analysis of the results, or publication in peer-reviewed scientific journals. In addition to the STTP, long-term scientifically based experiments are crucial in evaluating the long-term extrapolations that are required by PA analysis. These experiments are not part of the present experimental program; however, they should be an essential aspect of the iterative PA process, which will be required even during the operational phase of the repository. The needs of the AST program are not unique to the WIPP project. DOE inevitably will need a strong program in basic actinide chemistry as part of its nationwide environmental remediation effort. Prudent and efficient planning could consolidate the needs of the WIPP program within a broader program of basic research in actinide chemistry.

The interaction between principal investigators in the AST program and modelers in performance assessment has been disappointingly limited. Thus, it still is not clear how the AST data (experimental or theoretical) will be used in the performance assessment. The PA analysis has not yet provided the type of guidance to the experimental program that is required in order to determine when the work is, in fact, complete.

Concerning the use of performance assessment, there is still no clear distinction between the uncertainty associated with the selection of conceptual models in the AST and the uncertainty that results from the experimental data. This distinction is essential in evaluating the impact of the AST on PA analysis and conclusions.

The use of experimentally determined batch K_ds alone is not sufficient to justify the assumption of retardation of actinides during transport through the Culebra Dolomite. Field confirmation and/or a more fundamental understanding of sorption processes are needed.

Based on the current assessment of the WIPP site, the committee believes that the required work on the actinide source term (actinide concentrations in brine, colloid formation and transport, and nuclide retardation) can be performed simultaneously with the operation of the facility.

Chapter 6

Non-Salado Hydrology

The purpose of this chapter is to review and comment on the state of the U.S. Department of Energy's (DOE's) understanding of hydrology in formations at the WIPP site above the Salado. Studies of these subsurface aquifers ideally should aim to define the flow of ground water as well as possible. Virtually all feasible pathways by which radionuclides can move from the repository to the accessible environment depend on ground-water transport.

Site hydrology considerations are thus relevant to predicting the fate of any radionuclide release from a WIPP repository. Any such release will depend predominantly on three factors:

1. the volume of water (brine) that circulates through the waste-filled rooms and is released;[1]
2. the concentration of radionuclides in this brine; and
3. the rate of transport of the radionuclide-contaminated brine through the permeable sediments above the Salado Formation (e.g., Culebra Dolomite, Dewey Lake Red Beds) to the accessible environment.

WIPP repository design measures that can be taken to reduce the volume of brine release have been discussed in Chapter 4. Detailed studies to determine radionuclide solubilities, and hence concentrations, in brine have been discussed in Chapter 5. This chapter considers the third factor, considerations of ground-water flow and radionuclide transport.

If it can be demonstrated reliably that the concentration of radionuclides in the water and their rate of movement through the repository to the accessible environment together do not cause the release limits specified in 40 CFR 191 to be exceeded, then the repository can be considered to be in

compliance with these regulations. Such a demonstration would, in principle, involve a combination of the three factors enumerated above, with their relative importance dictated by the natural site characteristics and any engineered enhancements to the repository design.

REGIONAL HYDROGEOLOGIC MODELING

In the interests of brevity in this overall report, the significant advances in the understanding of the hydrogeologic framework that have been made since the WIPP investigations were started (i.e., since the mid-1970s) are not discussed in detail in this chapter. Appendix A summarizes the regional hydrogeology of the WIPP site.

The latest formal analysis available to the committee and discussed here is the 1992 Performance Assessment (PA). This model assumed a two-dimensional, steady state, horizontal flow of ground water in the Culebra Dolomite, with simple parameterizations of both chemical and physical retardation mechanisms to describe solute transport.

The discussion below outlines the main shortcomings of the 1992 PA models and analysis. Some of the criticisms reflect current limitations of the state of art of ground-water and transport study in general. Some criticisms can be resolved relatively easily, but other criticisms may require considerable effort to address.

The assumptions of the 1992 PA predicted a release of a mass of contaminants under disturbed conditions that, for a compliance demonstration, required a reliance on two features: long ground-water travel times in the Rustler Formation (primarily in the Culebra Dolomite) overlying the Salado Formation, and significant retardation of radionuclides. If this remains true in the PA calculation submitted as part of DOE's compliance certification submission in October 1996, then the committee believes that DOE will have to

---

[1]For 40 CFR 191 compliance, the total volume over 10,000 years is to be considered.

improve the case presented in the 1992 PA substantially to establish the technical credibility of the assumptions and values used in the analysis.

Most of the critiques in the following discussion of non-Salado hydrology relate to assumptions or numerical methods in the 1992 PA that need to be documented better or evaluated further before a high degree of reliance can be placed in the overall model results. For example, the PA models allow chemical retardation or matrix diffusion to attenuate radionuclides after release to the Culebra. Although it seems likely to the committee that some chemical retardation and matrix diffusion will occur, the documented basis for the assumptions about retardation in the 1992 PA is very weak. It is possible, and in some cases probable, that the final PA models and more recent experimental results—which have not yet been reviewed by the committee, but which will be included in the final compliance submission—will resolve many of these concerns.

Considerable effort has been expended by DOE over the last four years to increase its understanding of the ground-water and transport regimes at WIPP, and informal reports of significant progress have been made. The considerations raised in the following discussions point out shortcomings in past (i.e., 1992) modeling efforts, but the committee has determined that they do not challenge its position that DOE should be able to complete the work required to show that WIPP can comply with 40 CFR 191, and that the repository can probably be shown to present a low risk of human exposure to radionuclides, even in the event of a severe (E1E2) intrusion. However, to the extent that compliance demonstrations may rely on ground-water flow and transport in the non-Salado formations, they will require examination of at least some of the hydrogeology and transport issues discussed below.

FLOW AND TRANSPORT MODELING

A general discussion of ground-water modeling efforts and common difficulties is provided in this section (see Box 6.1 for a summary of key features). Supplemental background discussions of these issues are included in Appendix F. Models that simulate ground-water flow and transport processes are an essential tool in the characterization, analysis, and performance assessment of WIPP. Model predictions also lie at the heart of the PA process for the WIPP site (see Chapter 2). Therefore, it must be demonstrated that a high level of confidence in the PA models is justified.

The accuracy and precision of predictions of transport of radionuclides through a deep regional aquifer system depend on the level of understanding of the processes governing flow and transport through that system, how well the relevant properties of the system are defined, and how accurately the boundary conditions of the system are known. If the assessment of the isolation capability of the site is expected to provide a reasonable level of confidence that the predictions bound the range of plausible values, then the predictions must be based on an appropriate and adequate level of understanding of the processes, properties, and boundaries of the aquifer system and of how these factors may change in the future. If the numerical PA models represent the wrong processes, ignore relevant processes, or incorporate erroneous or biased assumptions or coefficients, then the results may be misleading. These considerations of ground-water modeling efforts as applied to WIPP are discussed below.

Sources of Modeling Error

In applying ground-water models to field problems, there are three primary sources of error, as described by Konikow and Bredehoeft (1993):

1. *Conceptual errors:* theoretical misconceptions about the basic processes that are incorporated in the model.

2. *Numerical errors:* the difference between a true and exact solution to the governing equations and the solution developed by the equation-solving algorithm.

3. *Uncertainties and inadequacies in the input data:* errors arising from the inability to describe actual aquifer properties, stresses, and boundaries. These include measurement errors, interpolation and extrapolation errors, and inaccurate predictions of future conditions.

BOX 6.1 Flow and Transport Modeling

Modeling of transport in this regional flow system requires the following:

1. An adequate conceptual model of the flow system, including

  - water table and potentiometric elevations
  - definition of major hydrostratigraphic units
  - description of potential preferential flow paths, such as fractures
  - evaluation of temporal changes (whether or not system is at steady state)

2. For significant hydrostratigraphic units, descriptions of

  - flow and transport mechanisms
  - flow and transport properties
  - nature and rates of chemical reactions, such as sorption, to estimate chemical retardation

3. Mathematical models that

  - incorporate the relevant processes
  - are not subject to large numerical errors

In most model applications, conceptualization problems and uncertainty concerning the data are the most common sources of error. All three aspects of error are discussed below with reference to the PA models for WIPP.

Conceptual Errors

If the conceptual models upon which the PA models are based are in error, then all of the complementary cumulative distribution functions (CCDFs; see Chapter 2 and Appendix B) and risk assessments derived from the PA also may be in error. Although the tool of PA is designed to account for uncertainties in the parameters, this assessment of uncertainty is necessarily within the context of the PA model used. Factors and processes that are not incorporated into this model are conceptual errors that are not treated by the PA sensitivity analysis.

The 1992 PA for WIPP did not assess adequately the uncertainties in the underlying conceptual model, which might be a more significant effect than the uncertainties in the parameters. For example, because the 1992 PA ground-water model was two dimensional, it cannot be used directly to evaluate the importance of a three-dimensional flow. Errors in a conceptual model are often difficult to detect, and sometimes field measurements can be reproduced by several alternative conceptual models. The alternative models may make little difference over the short duration of a field test but may make a large difference over the life of the repository and predictive horizon. It is therefore critical that the PA conceptual models be rigorously and independently tested.

There also must be consistency among the various conceptual models that describe different elements of the subsurface ground-water system. For example, hydrologic concepts indicate that ground water in the Culebra Dolomite at and near the WIPP site flows generally from north to south. Chapman (1988), among others, notes that there are significant decreases in salinity and changes in water quality in the direction of flow. Siegel and Anderholm (1994, p. 2299) state that "no plausible geochemical process has been identified that would cause this transformation in a hydrologically confined unit." Thus, geochemical interpretations appear to conflict with hydrologic understanding.[3] Such a discrepancy reflects a lack of comprehensive understanding of the subsurface system, which opens the possibility that some aspects of the overall conceptual model of the Culebra are flawed, and that predictions based on this model are likewise uncertain.

Numerical Errors

The 1992 PA model used the SECO ground-water modeling computer codes to simulate flow and transport. The SECO codes have been developed by recognized experts in numerical methods, and the transport code apparently represents a major advance in controlling and limiting numerical dispersion. However, because the SECO codes are not yet used widely, some "bugs" and other problems in the use of the codes may still remain undetected. Experience with other ground-water model codes indicates that code developers often cannot comprehensively test the code for all possible ranges and combinations of parameter values. Model applications by a variety of users and organizations to a range of problem types and hydrogeologic conditions often will result in the discovery of code deficiencies that were not detected by the model developers, regardless of how comprehensive their testing efforts.

The SECO codes have not yet been subject to extensive benchmarking or international comparisons with other widely accepted codes, such as have been performed for other codes under the recent International Hydrologic Code Intercomparison Project (HYDROCOIN) and the International Project to Study Validation of Geosphere Transport Models (INTRAVAL; Swedish Nuclear Power Inspectorate, 1987, 1990). The committee believes that it is inappropriate to rely exclusively on these relatively new codes before such benchmarking evaluations have been completed in an open, broad-based scientific forum to yield more confidence in the codes.

---

[3]Possible explanations are that prehistoric flow was in a different direction than current flow, or that geochemical or hydraulic interpretations are flawed, or that vertical components of past or present flows exist that have not been accounted for.

Data Uncertainty

The PA models make predictions based on ranges of values of parameters. The 1992 PA assumed that most of the parameters in the model are "independent random variables even though it is known that some were dependent on others" (Sandia National Laboratories, 1992). It is further stated (Sandia National Laboratories, 1992, p. 1-21) that "the effects of neglecting correlations on the sensitivity/uncertainty analyses are generally unknown." It may be a reasonable first approximation to assume independence of parameters in sampling (i.e., to ignore interactions between variables and their codependence on other factors). However, such an assumption must be carefully tested, and refined if necessary, before a final PA is completed. Independent sampling can yield simulations based on implausible (or contradictory) combinations of parameters. If results of these simulations yield low-risk outcomes, then including the results of such simulations in the generation of CCDFs might skew or bias statistical distributions of the results.

The committee recommends that the assumed independence, ranges, and distributions for the various parameters that are sampled in PA analysis be examined closely for reasonableness and consistency with observed data.

WIPP 1992 PA MODEL ANALYSIS

Regional Ground-Water Flow

As noted previously, the 1992 PA models assumed that ground-water flow through the Culebra Dolomite was two-dimensional, steady-state, and horizontal. Investigations of the larger three-dimensional regional system, of which the Culebra is actually a part, have begun relatively recently but have not yet been completed. These investigations include a new synthesis of all the regional hydrogeologic data and will lead to the development of a three-dimensional numerical flow model. This more comprehensive analysis of the ground-water flow system is very important and relevant to the larger issue of conceptual uncertainty. Among other benefits, this type of analysis may help to resolve existing discrepancies between the geochemical data and the present conceptual model of the flow system. Until the results of these ongoing studies have been completed and their impact on assumptions about the Culebra has been evaluated, it is too early to determine whether or not many of the simplifying assumptions incorporated into the 1992 PA were reasonable.

Ground-water flow and solute transport are affected strongly by heterogeneity in aquifer properties. Fractures represent an extreme but prevalent type of heterogeneity, so their characterization is important for reliable model development. However, the properties, continuity, and extent of fractures in the Culebra are still highly uncertain. One cost-effective approach to characterize fractures is to complete detailed mapping of any surficial expressions of fracture traces and lineaments at both local and regional scales. This should have been completed in the early stages of site characterization, but there is no written documentation of such analyses having been done.

Fracture and lineament mapping have been applied successfully to several potential fractured-rock repository sites (e.g., for applications in Sweden, see Wikberg et al., 1991; Ahlbom et al., 1992; Swedish Nuclear Fuel and Waste Management Company, 1995). Surface and subsurface features that might represent fracture zones and lineaments were defined by using remote-sensing methods (such as aerial photography and satellite imagery), topographic data (including computer analysis of digital elevation data), and several geophysical surveying techniques (including geomagnetic, electrical, and seismic methods and some recently developed high-resolution, three-dimensional tomographic techniques).

These approaches have helped to clarify the possible existence and spacing of major through-going fracture zones. Although the lithology at WIPP is different from that at the Swedish sites, these approaches have worked in other sedimentary rock basins, and the committee believes that they should be applied in the characterization of the WIPP site. An approach to fracture definition that is potentially more definitive, but also significantly more expensive, would involve drilling of slanted or even horizontal boreholes. However, because this probably would require an extremely expensive and time consuming operation, the possible benefits should be carefully weighed against the likely costs before a decision to undertake such a study is made.

Confidence in PA Analysis of Regional Ground-Water Flow

Significant accomplishments in site characterization have advanced the level of understanding of the regional ground-water flow system at the WIPP site during the past 20 years (see Appendix A for a summary discussion). However, the degree and comprehensiveness of the understanding of some aspects of the local and regional ground-water flow systems are lower than desirable relative to the importance of the repository and to the time and money spent studying the WIPP site and constructing the repository. The following points support this contention:

• There is a disparity (as discussed above) between the regional flow system as determined by geochemical and isotopic interpretations and that indicated by hydrologic analyses. This disparity reflects an inadequate level of understanding of the whole system, and it should be resolved.

• The water table in the region, which represents the top of the saturated zone, has not been defined. This important boundary should be defined, or a reasonable explanation of either why it is not important or why it is technically or economically infeasible to determine should be documented.

• The rates, locations, and mechanisms of natural recharge to and discharge from the Culebra Dolomite are poorly defined (see Brinster, 1991).

• Apparently, there have been no detailed hydrologic studies of the unsaturated zone at the site. Such studies would help to (1) estimate if, why, and how much recharge may exist in the vicinity of the site; (2) define the water-table position; (3) predetermine an appropriate response should there be a spill or accident on the land surface at the WIPP site; and (4) predict how much and where recharge might increase under a scenario of climate change that yields increased precipitation.

• An adequate explanation is lacking for observed changes in water levels in the Culebra, where trends of rising water levels have persisted for several years.

Observed changes in water levels from assumed steady-state conditions were not incorporated into the 1992 PA analysis. However, if the causes of the observed water-level changes during the last several years are unknown, then how is it to be known that even greater changes in the flow field might not occur in the near future? Such changes might invalidate the PA assumptions and predictions.

• These water-level changes are affecting hydraulic gradients, flow directions, and flow velocities and thereby provide some evidence against the assumption that steady-state flow prevails in the aquifer, an assumption that is the basis of many WIPP hydrologic analyses. Small "violations" of this assumption probably would not affect the conclusions to a noticeable degree. However, if the trends observed over the past five or six years persist, recur, or become more widespread, flow systems might change enough to negate the PA analyses based on the assumed steady-state flow field.

• There has been too little accounting for the three-dimensional nature of ground-water flow and consequent leakage through confining layers. If vertical components of flow do materially affect the Culebra, then the ground-water models of the Culebra that were calibrated under the assumption that flow is horizontally two dimensional must include compensating errors to offset ignoring the vertical leakage. Such errors can induce a bias in long-term predictions of flow and transport.

Some necessary and feasible experiments and analyses (of the kind mentioned here), which would lead to a more comprehensive understanding of the ground-water flow system and to more reliable PA models, have either not been undertaken or are not completed yet.

Ideally, the regional system should be understood with sufficient accuracy and precision that professional hydrologists will have confidence in the analyses and predictions, will believe that additional data collection is not critical, and will be convinced that there are no major "surprises" left in the characterization of the system.

Although it is difficult to define precisely when a requisite level of confidence in the hydrogeologic analyses has been achieved, the committee does not believe it has been reached in the 1992 PA. Too many questions remain about the models, assumptions, and data used to analyze the cases in which a significant release of contaminants to the Culebra or other formations is projected to occur.

The potential weaknesses in the ground-water models will be important to a demonstration of compliance only if there are significant releases from WIPP to the stratigraphic units shallower than the Salado. The weaknesses in hydrogeologic modeling discussed in this chapter should be balanced against an examination of the assumptions from which the estimated releases are derived, and assumptions concerning pathways to cattle that drink from stockwells tapping the Culebra. For example, even if contaminants were transported through fast-pathway fractures in the Culebra and were not subject to any retardation processes, the very same conditions that would enable this to occur would assure that the water carrying the contaminants would be far too saline for cattle to consume. This would appear to eliminate the potential for future human exposure, which is an underlying basis for the overall concerns about radionuclide releases.

The non-potable saline character of the ground water in the Culebra will greatly reduce the exposure hazard even though the radionuclides released to the accessible environment may exceed the limits required for compliance in 40 CFR 191. Should the 1996 PA analysis be able to demonstrate convincingly that the releases are sufficiently low to meet compliance standards, then the exposure hazard would be correspondingly lower.

Solute Transport

Transport of radionuclides released into an active ground-water system can and may occur at a rate less than the rate of the ground-water flow, due to physical and chemical interactions with neighboring rock. This radionuclide retardation, where it can be shown to exist, can add important additional assurance with respect to the radionuclide containment capability of the repository. Hence, studies of the radionuclide retardation properties of water-conducting rock formations at WIPP (see Appendix A) are an important complement to the studies of ground-water flow.

Following this discussion of solute transport, the specific retardation mechanisms are treated in turn.

The solute-transport model of the Culebra Dolomite is based on a theoretical understanding of transport and dispersion processes in ground water. However, this underlying theory is itself somewhat weak because the scientific understanding of transport processes in heterogeneous media, particularly in fractured rock systems, is still in a stage of relatively rapid development and evolution. Much remains to be learned, and much ongoing research is devoted to developing a better understanding of the governing processes and parameters at large scales in complex field environments. Recent research indicates that many transport processes and parameters are highly scale dependent. This is one reason why there is often little basis for extrapolation of theory and lab tests to the field environment for predictive purposes.

The 1992 PA separately sampled and independently varied porosity, fracture spacing, and transmissivity. However, it is known that correlations exist among these parameters. Thus, it seems likely that the existing approach may yield unreasonable or unlikely combinations of these parameters in some of the individual simulations. The net impact on the overall PA is uncertain, but the committee believes that the parameter sampling procedure should be examined closely and refined if necessary.

The PA models represent leakage from a borehole into the Culebra as an initial concentration of solute in the transport model for a steady-state flow field. However, transient changes in the flow field, which are induced by the volume of water carrying the radionuclides into the Culebra, are not considered. While the leak is occurring, the velocity away from the well would be greater than the velocity under steady-state conditions, and some of the radionuclides would migrate faster and further than represented in the PA model. The committee recommends that the PA process carefully test and demonstrate that the steady-flow assumption and simplification, which ignore the initial fast advection, are reasonable and do not introduce significant errors.

Confidence in PA Analysis of Solute Transport

The effective porosity (ε) is a critical transport parameter for which few measurements are available. If one assumes that $\varepsilon = 0.001$, which is representative of fracture flow, then travel times to the regulatory boundary would be on the order of hundreds of years. However, if $\varepsilon = 0.16$, which is representative of the rock matrix, travel times would be on the order of many thousands to tens of thousands of years. This could mean the difference between compliance and noncompliance of the site.

The 1992 PA assumed that porosity and transmissivity vary independently. This is a potential weakness in the analysis that needs to be tested, because physically reasonable arguments lead to correlations between the two variables. In this regard, a key issue for the PA flow model is whether the relation between transmissivity and porosity is in fact independent, or if the relation is direct or inverse. Either relation could be conceptualized (or perhaps rationalized), based on present knowledge of the Culebra Dolomite. However, any consistent relation would be contrary to the WIPP Project PA procedure of sampling independently from the populations (or distributions) of transmissivity and porosity. Thus, many simulations would be based on unlikely or contradictory combinations of parameter values. This could bias the PA model-generated statistics on which the CCDFs are based.

The PA sampling approach has yielded some individual outcomes to date that have exceeded the compliance standard. These particular simulations should be examined carefully to highlight and document the combinations of parameter values and boundary conditions that led to prediction of an unacceptable release of radionuclides to the environment. This evaluation would allow analysts and reviewers to focus on an assessment of the reasonableness of those specific and critical conditions, and would serve as a quality assurance check on the statistics generated by the PA analyses.

Some of the 1992 PA simulations of solute transport have allowed physical and/or chemical retardation, as would be derived from matrix diffusion and chemical sorption, respectively. To date, however, no reliable and representative field measurements of the parameters controlling matrix diffusion and chemical retardation have been carried out at the WIPP site.

Physical Retardation Via Matrix Diffusion

In the 1992 PA model of the Culebra Dolomite, physical retardation may occur in response to matrix diffusion in a dual-porosity system. The magnitude of the retardation is proportional to the effective surface area and volume of the blocks adjacent to the fractures through which advection is occurring. If the contaminant plume spreads through a larger volume of the aquifer, then matrix diffusion will be more effective in removing contaminants from the active flow field. If the transport model artificially spreads the contaminant plume into a larger volume of the aquifer than would actually occur, then the model will overestimate the retardation caused by matrix diffusion, even if the diffusion parameters themselves are estimated accurately.

The artificial spreading of contaminants can arise from errors in the numerical solution, from errors in the initial conditions, and from errors in the conceptual model. All are present to some degree in the WIPP PA model. The committee recommends that PA assumptions and analyses for cases including matrix diffusion be supported by additional model studies to demonstrate that the analyses do not overestimate the magnitude of matrix diffusion because of artificial spreading of the plume. This can be accomplished, for example, by applying a transient ground-water flow model that uses, for a relatively short time period, a much finer spatial grid and smaller time steps near the area close to a human intrusion (HI) borehole. These results can then be compared with those of the coarser PA model for equivalent travel distances and elapsed times.

Confidence in PA Analysis of Physical Retardation

If PA results indicate that repository compliance with EPA standards depends strongly on matrix diffusion, more reliable field evidence to support the assumed range of coefficients should be obtained. In this case, field tests should be completed at several different locations in the vicinity of the WIPP site because of the spatial variability in the properties of the system.

Chemical Retardation Via Reactions

The PA models combine all of the complex chemical reaction processes, such as sorption or exchange, that might retard the movement of a particular chemical species relative to the movement of a nonreactive solute into a single simple parameter—the retardation factor R_f. This factor, in turn, is related to a distribution coefficient K_d. Although the concept of a retardation factor is computationally efficient, it inherently assumes that reaction processes are linear, instantaneous, reversible, and homogeneous.

The concept generally works well in chemical plants where chemical engineers apply it to fluid flow and solute transport through tanks filled with relatively homogeneous, reactive, porous media. However, in a complex regional aquifer system, the reactions rarely are subject to these ideal conditions. The theoretical basis for applying a retardation factor to a field environment is therefore weak, at best, and its application to large-scale hydrogeologic settings, without restrictions or qualifications, has been criticized in the literature (e.g., Reardon, 1981; Miller and Benson, 1983; Valocchi, 1984).

From the perspective of a compliance assessment, use of the simplifying retardation factor may either underestimate or overestimate the overall amount of retardation of a solute relative to the flow of water, depending on the particular chemical species and the types and rates of reactions that affect its fate and transport. To probe this issue, the nature of the expected isotherm for each solute of concern (and within the range of expected concentrations) should be determined experimentally.

Scientists from Sandia National Laboratories recognize the numerous difficulties and weaknesses in applying a linear sorption model for predicting radionuclide transport through the Culebra Dolomite, and many of these concerns have been documented (Novak, 1992). As an example of one of several serious concerns, Novak (1992) points out that available K_d data for uranium sorption in the Culebra are very sensitive to water composition and vary by more than three orders of magnitude for selected representative compositions of Culebra brine. Novak (1992, p. 55) states that "existing radionuclide K_d data are insufficient for predicting radionuclide migration in the Culebra with a sound scientific basis within reasonable certainty." Ongoing experiments will help to resolve this uncertainty, but results and analyses are not yet available for review.

Confidence in PA Analysis of Chemical Retardation

If performance assessment results indicate that compliance is contingent on the occurrence of chemical retardation, it will be necessary to provide more evidence that the PA approach is reasonable for these conditions. The committee believes that the use of sorbing tracer tests in the field would be highly beneficial for building confidence in the analysis and approach to chemical retardation, in addition to determining whether or not the conceptual model is applicable to the field environment. Because the specific solutes of interest here, such as plutonium and americium, cannot be used in tracer tests, reasonably appropriate surrogate tracers must be used, or field evidence of sorptive behavior must be ascertained from analysis of previous accidental releases elsewhere.

As hypothesized for flow parameters for the Culebra Dolomite, a correlation may also exist between K_d (or R_f) and either transmissivity or porosity. For example, relatively high transmissivity values might coincide with zones of low or zero clay content. Because clay probably yields a higher sorptive capacity per unit volume than dolomite, where transmissivity and flow velocity are the highest, there may be very little chemical retardation. This illustrates one type of predictive error that might be associated with applying one average effective R_f to the entire ground-water system. The greatest risk of release in that case is related to flow through transmissive fractures having a lower than average R_f.

It is unclear whether the PA approach to sampling values of R_f is appropriate if the chemical retardation in critical high-transmissivity channels is not represented accurately by the average R_f assumed for the entire aquifer system. Again, the only way to address this issue in a manner that would instill confidence in the predictive analyses is through a series of in-situ tracer tests or other field-based analyses.

DISCUSSION AND SUMMARY OF FINDINGS, CONCLUSIONS, AND RECOMMENDATIONS

This chapter and Appendix F have focused on potential sources of error in the preliminary PA modeling of the Culebra Dolomite. Therefore, the reader should not infer from the tone of this chapter that the WIPP ground-water models are necessarily inaccurate or unacceptable. Rather, the intent is to indicate where the preliminary analyses presented in the 1992 PA need to be revised, strengthened, or better documented. Because there have been four years of additional studies since the 1992 PA, it is possible that recent experimental results and the final PA models, which have not yet been reviewed by the committee, will have resolved many of these questions.

A key question is whether the uncertainties in the properties of the subsurface system above the Salado Formation need to be resolved (or minimized) further to demonstrate adequately the integrity of the WIPP site with respect to containment of radionuclides. Alternatively, is enough information about the subsurface system already available to predict that any potential release from the proposed repository will not cause radionuclides to migrate faster or further than is acceptable from a regulatory standpoint?

These issues are addressed separately for the undisturbed case and the disturbed (i.e., HI) cases described in Chapter 2.

Undisturbed Repository

The committee finds that it is reasonably certain that the WIPP repository will provide an adequate level of long-term isolation of TRU waste under a broad range of natural conditions and stresses for tens of thousands of years into the future. Provided that seals and plugs (as discussed in Chapter 4) are effective, there is no evidence to indicate that any brine that might be contained in the repository after it is closed is likely to leak through the Salado Formation and be released into the regional ground-water flow system in shallower formations. Nor would the committee expect that ground water flowing above or below the Salado might dissolve the salt or in any other way come in contact with the wastes that would be stored in the repository.

For consideration of the undisturbed case, the present state of understanding of the ground-water flow system above the Salado is adequate. The committee finds that the overall permeability and creep properties of the Salado are such that, with effective repository seals and without some form of human intrusion, the flow of radionuclide-contaminated brine from the Salado is very unlikely. Further studies of the hydrogeology of non-Salado units are therefore not necessary to support compliance for the undisturbed case.

Summary of Undisturbed Case

Provided that plugs and seals in the Salado Formation are effective (see Chapter 4), there is high confidence in the conclusion that the WIPP repository will isolate TRU waste adequately from the environment for many thousands of years if it remains undisturbed by future societies. For this undisturbed case, continued studies of non-Salado hydrogeology are not necessary.

Disturbed Repository

In the WIPP PA (see Chapter 2), the HI scenario assumes that future drilling will facilitate upward leakage of brine from the repository through a borehole, thereby permitting radionuclides to be released into an overlying geologic formation through which lateral migration offers a possible pathway for release of radionuclides to the accessible environment. If releases exceed regulatory limits, then a predicted rate of transport in the regional ground-water system above the Salado Formation becomes a critical factor in estimating cumulative releases.

The analysis of the disturbed case hinges on predicting flow and transport through rock formations above the Salado (see Appendix A). The issue of concern to the committee is whether the 1992 PA was based on an adequate conceptual understanding of the regional ground-water flow system near WIPP and whether the relevant parameters had been quantified adequately to support either "realistic" or "bounding" models in the PA.

For example, the 1992 PA assumed that the most likely pathway for a lateral release of radionuclides is through the Culebra Dolomite. However, a release to the shallower Dewey Lake Red Beds also is possible and does not appear to have been considered or evaluated adequately. Releases of saline ground water from leaking deep boreholes directly to the water table are known to have occurred historically in other areas (for example, see Van der Leeden et al., 1975). The Dewey Lake is a plausible exposure pathway because it is known to contain some potable water. Among other concerns, climate changes can increase the recharge rate and raise water levels. In the committee's opinion, releases to the Dewey Lake cannot be discounted summarily; if a borehole to the Salado or to the Castile Formation were to connect these formations to brine at a pressure near lithostatic, then the hydraulic gradient (driving Darcy flow) would be sufficient to enable leakage into both the Culebra and the Dewey Lake if a pathway to either formation were to exist.

Also, the amount and rate of brine release appear to be very sensitive to the assumed permeability of the material that is filling the leaky borehole. In light of the importance of that parameter, values for borehole permeability that are used must be realistic and technically defensible.

If the final analyses of the disturbed-case scenarios yield projected releases of contaminants to formations above the Salado that are less than the release limits of 40 CFR 191 (i.e., for which compliance results even if the releases moved across the regulatory boundary in less than 10,000 years), then no further hydrogeologic characterization work is needed. This means that compliance could be demonstrated even if one assumes that no retardation occurs and that ground-water flow and transport occur through fractures.

Summary of Disturbed Case

For the disturbed case, if the mass of contaminants and brine released is as great as assumed in the 1992 PA, then the ability of the site to isolate waste satisfactorily under such "disturbed" conditions has not yet been demonstrated adequately. However, such a demonstration appears to be a plausible outcome if the following problems, issues, or limitations of present models and data are resolved through further studies.

1. The PA approach must assess the sensitivity of the predictions to uncertainties in the underlying conceptual

models. Alternative conceptual models are feasible, and those that are plausible and that could lead to significant releases should be evaluated. Completion of ongoing three-dimensional analyses of the regional ground-water flow system is critical to this effort and should be given a high priority.

2. The non-Salado site characterization efforts and PA models have been overly focused on a two-dimensional flow analysis of the Culebra Dolomite. The minimal levels of consideration and evaluation of other potential hydrogeologic release pathways (such as the Dewey Lake Red Beds) and of the interconnection between the Culebra and adjacent formations have not been justified adequately.

3. Fractures are a dominant control on the transmissivity of the aquifer system at the WIPP site and represent the highest-velocity channels for migration of contaminants. A better definition and understanding of the nature, density, spacing, length, and interconnectedness of fractures and fracture networks are critically needed. As one step to help achieve this understanding, the committee recommends that modern topographic and geophysical methods of analysis be applied in the site characterization efforts.

4. In general, the types of chemical and geochemical reactions, rates of reactions, and their dependence on spatially heterogeneous mineralogy and changing aqueous geochemical parameters (e.g., salinity, pH, and Eh) are not well characterized for the HI scenario at the WIPP site. Representing all of these sources of variation in the PA model as a single parameter (i.e., an effective retardation factor), as was done in the 1992 PA, has a very weak scientific basis. If the PA assumes that chemical retardation occurs in the Culebra, then supporting evidence from the field environment should be available before any predictions based on that assumption are accepted.

5. Similarly, if matrix diffusion is to be relied upon in the PA models to demonstrate compliance, then additional controlled field experiments must be completed to define rates of diffusion more accurately and to assess spatial variability in the process and parameters. Furthermore, additional testing and documentation are needed to eliminate concern that the net retardation caused by matrix diffusion is being calculated inaccurately in the numerical models.

6. For the disturbed case, if it can be shown that the mass of contaminants and brine released would be significantly less than previously estimated (as might be derived, for example, due to changes in the estimated severity of human intrusion, refinements in estimates of the actinide source term, or in the effects of borehole closure), then further studies of the Culebra would not be necessary. That is, if the total amount of solute entering the non-Salado formations is determined to be sufficiently small to assure compliance, then studies of ground-water transport are not needed for this purpose.

PERSPECTIVE BASED ON INTERNATIONAL REPOSITORY SITING EFFORTS

Insight into the difficulties of hydrogeologic site characterization issues is offered based on examples from and comparisons with international siting efforts. In efforts to site geologic repositories in other countries, a good quantitative understanding of the regional ground-water flow regime and the influence of the geological environment of the repository on radionuclide transport are often the most critical factors needed to determine whether or not a potential repository site can provide adequate long-term isolation of the waste. This is usually true for both the undisturbed and the disturbed cases.

Several countries have opted to site repositories in rock masses of low permeability, such as granites, in regions of low hydraulic gradient, so that flow rates will be very small. Determination of the rock-mass permeability in these cases is often complicated by the presence, on the field scale, of fractures that are not usually present or tested in laboratory specimens. Fractures also complicate larger-scale tests conducted directly in the field, making it difficult to extrapolate to the regional scale. Fractures tend to follow a log-normal distribution in which well-developed, extensive, and often highly conductive fractures are spaced widely apart, with smaller, less extensive fractures becoming progressively more abundant. On the smaller scales, these are not always interconnected and hence may be nonconductive. Definition of rock-mass permeability of fractured systems is difficult, and factors that influence radionuclide transport processes are at least as complex.

Predicting regional ground-water flow and radionuclide transport behavior with the reliability and detail required for repository performance assessment is a complex and formidable challenge. It has been a main focus of international research related to repository siting and design for more than 25 years. Although substantial advances have been made—with project scientists and consultants making notable contributions—significant problems remain. Each potential repository site has unique features. To the extent that long ground-water travel times and significant physical and chemical retardation are necessary to ensure adequate isolation of radionuclides from the accessible environment, characterization of the site is correspondingly necessary to address these problems.

Most repository sites worldwide are located in rock that is part of an active regional ground-water regime. In these cases, except for engineered barriers in the immediate vicinity of the waste, the repository is physically close to circulating ground water. By contrast, locating WIPP centrally within the 400-m-thick, essentially impermeable, Salado bedded salt deposit, some 200 m below the permeable, saturated sedimentary Rustler Formation, which consists of fractured limestones and dolomites, is a major design advantage of this repository. The restrictions to ground-water transport provided by the Salado Formation provide a valuable "buffer zone" between the repository and the non-Salado permeable formations above it, to which the site hydrology considerations of this chapter apply.

Chapter 7

Perspectives

The NRC Committee on the WIPP has followed the WIPP project from its infancy in 1978. There have been some changes in committee membership, but there has been sufficient continuity that the committee has remained aware of the significant scientific and technological issues, as well as the associated policy and budgetary constraints on the WIPP project. This report focuses on the committee's understanding of the specific scientific and technological issues that form the core of DOE's compliance application to the U.S. Environmental Protection Agency and their influence on the perceived ability of WIPP to isolate TRU waste from the biosphere.

During its deliberations, the committee also has arrived at several general conclusions with respect to the overall integrity of WIPP, the difficult problem of human intrusion, and the role of Performance Assessment. Because of their importance to this report and the overall integrity of WIPP, the committee provides a general discussion of these issues in this chapter.

INTEGRITY OF THE WASTE ISOLATION SYSTEM

A previous National Research Council report provides a useful introduction to the topic of geologic isolation at WIPP. The report states that the objective of a geologic waste isolation system is

to protect humans now and in the future by isolating the waste from the environment effectively enough and for a period of time long enough that the amount of radioactive material ever reaching the biosphere will present no unacceptable hazard. To achieve this objective, a hierarchy of mechanisms exists to reduce the release of radionuclides to the biosphere and thus have the waste isolation system meet the

criterion of overall performance. One or more of the following release control mechanisms must be sufficient to meet the panel's criterion:

- *delay of ingress of water*
- *slow dissolution of radionuclides*
- *slow release from the waste package*
- *long groundwater travel time*
- *delay due to sorption in the geologic medium*
- *dispersion*
- *dilution*

The system must guard against an unacceptable release of the radioactive material into groundwater and the transport of this contaminated water to the biosphere—a principal pathway by which some portion of the buried radioactive material may eventually reach humans. (NRC, 1983, pp. 3-4)

Although the 1983 report gives primary emphasis to geologic isolation for high-level, spent reactor fuel or reprocessed waste (hence the high ranking given above to "slow release from the waste package"), the list of important criteria is essentially the same for a transuranic (TRU) waste repository. The total radioactivity per unit volume of TRU waste planned for WIPP is roughly on the order of one-thousandth of the activity for an equal volume of spent fuel waste, but many of the same long-lived radioactive isotopes are present, and so geologic isolation is necessary.

In the committee's opinion, location of the WIPP repository within the 600-m-thick Salado Formation, some 400 m below the overlying, permeable water-saturated sediments is an exceptionally robust way in which to delay ingress of water to the repository or, equivalently, egress of radionuclides from the repository to the accessible biosphere. Natural salt creep will result in the encapsulation of waste within what becomes essentially intact salt a few hundred years at most after emplacement. Sealing of the shafts with crushed and compacted salt also should ensure that, by natural salt creep processes, these potential conduits to the biosphere are sealed effectively. Thus, the waste should be sealed, permanently, within the Salado Formation, which has been stable for over 200 million years.

The committee is confident in asserting that, provided the repository is sealed effectively and remains undisturbed by human activity after final closure, WIPP will isolate TRU waste for long enough (i.e., many tens of thousands of years) to ensure that the radioactivity in the waste will have decayed to very low levels.

HUMAN INTRUSION

From the point of view of waste isolation, salt does have one drawback—it is frequently found in association with mineral resources, especially oil and gas. Drilling for these resources could result in one or more holes penetrating the repository and would provide a pathway to bring the waste into direct contact with ground water and the accessible biosphere.

An intrusive drilling event could remove the isolation from ground water provided by the 200-m-thick envelope of salt. Because dispersion and dilution are not significant factors at WIPP, the two remaining mechanisms to control radionuclide releases noted in the NRC (1983) report would be (1) slow dissolution of radionuclides, and (2) delay due to sorption in the geologic medium. Although WIPP does not include a "waste package," the backfill and seals around each waste room and panel serve a similar function. In fact, the salt "package" does not have the limited lifetime of a high-level waste canister. For WIPP, assurance of protection from excessive release of radionuclides due to drilling intrusion relies on (1) engineering efforts to improve repository performance (see Chapter 4); (2) improved understanding of the actinide source term (see Chapter 5); and (3) understanding of non-Salado Hydrology (see Chapter 6)—the same three mechanisms that are identified in the NRC (1983) report.

ROLE OF PERFORMANCE ASSESSMENT

What contribution can be expected from each of these component factors, and how much is needed, to ensure effective long-term isolation of the waste?

Total System Performance Assessment, or PA, is the method that has been used at WIPP to identify and examine all credible pathways by which radionuclides could be released to the biosphere from the repository and to estimate the contribution of each component of the pathway. Although both the undisturbed and disturbed (by drilling) cases have been examined, the DOE PA team has placed appropriate emphasis on the more complex and critical disturbed case.

The latest "complete" performance assessment available to the committee was published in 1992. The 1992 PA identified the same three key factors noted above, and its results suggest that WIPP could comply with the EPA release standards if certain levels of performance were attained from each component.

As noted principally in Chapters 2, 5, and 6, the committee has identified several significant deficiencies in the 1992 PA. These include the following:

1. The assumed permeabilities governing the flow of brine through the underground excavations were much too high. As a result, predicted flow through the repository due to an E1E2 intrusion (see Chapter 2, Figure 2.5) greatly overestimated brine flow, and hence radionuclide release.

In the 1992 PA, no credit was given for the ability of the excavations, especially when back-filled with crushed salt, to close within the order of a few hundred years, or for the potential of room and/or panel engineered seals and various other available engineering techniques to reduce both the consequences and the probability of serious drilling intrusions. For example, if the values for repository permeability used in the assessment are reduced progressively towards the value of intact salt (10^{-23} m^2) over 10,000 years, rather than maintained at a constant 10^{-15} m^2, the total predicted brine flow and radionuclide release would be much lower than indicated in the 1992 PA.

2. There was insufficient experimental justification for the assumptions made with respect to actinide

solubility, colloidal transport, and ground-water flow/radionuclide transport in the Rustler Formation overlying the Salado Formation.

Experimental studies to improve the basis for estimating actinide solubility and Rustler Formation hydrology and transport have now started, but these will require several years of additional work in order to arrive at reliable results. At this stage—with the results of the 1996 PA not yet available, and important experimental studies still in progress—the committee does not know how the results will affect predicted releases and compliance. It does appear to the committee that engineered design features to reduce the adverse consequences of an intrusion are available, but not all of these features may be cost effective or needed if low actinide solubilities and/or sufficiently long travel times and radionuclide retardation can be demonstrated by the PA. Thus, to inform decisions on implementing costly engineering design features, it is important that these experimental studies be continued to completion.

3. The 1992 PA was directed at demonstrating compliance with the radionuclide release limits stipulated in 40 CFR 191. No analyses were made of the health risks associated with these releases. At WIPP, radionuclides would be transported in brine that is considered to be, to a large extent, unpotable even to cattle. Consequently, the health risk will be very much smaller than if the water were potable.

In considering the radionuclide exposure hazard of WIPP to humans (see also Box 1.2), the committee recognizes the unpotable nature of much of the water in the Rustler above the Salado. The committee believes that accounting for the unpotable nature of the water could reduce the health risk by one or more orders of magnitude compared to the same release into a freshwater aquifer that is similar in size and has similar regional flow characteristics.

Taking into account all of the above considerations, the committee is confident in its judgment that DOE should be able to demonstrate that radionuclide releases at WIPP will be within the limits allowed by EPA, for both the undisturbed and disturbed cases, even with the severe criteria defined in 40 CFR 194. The associated health risks are likely to be well below the levels allowed under international standards. The results of

the 1996 PA are awaited with interest as a test of this judgment.

The heavy emphasis placed in the 1992 PA on the analysis of compliance with 40 CFR 191 for the "disturbed case"—that is, assessing the consequences of human intrusion—is appropriate in the committee's judgment because of the added complexity of the radionuclide pathways introduced by intrusion. Even so, it is important to recognize the remarkable ability of the salt at WIPP to isolate waste under undisturbed conditions.

THE NATURE AND FREQUENCY OF HUMAN INTRUSION

Although the consequences of some form of human intrusion should be assessed, it is also evident to the committee that there is no scientific justification for estimating the precise nature or frequency of such intrusions over the next 10,000 years. For compliance purposes at WIPP, EPA has decided that the frequency of drilling for oil and gas exploration and production during the past century will continue unchanged for the next 10,000 years, even though it is acknowledged that the oil and gas reserves near WIPP will be practically exhausted within the near future (i.e., in less than 50 to 100 years).

The EPA assumes that today's drilling activities are representative of the type of drilling activities that will occur in the future. Although the resources drilled for today may not resemble those drilled for in the future, the EPA

believes it is reasonable for the average rate to be projected over the next 10,000 years, based on the assumption that while oil and natural gas may be depleted, other resources (which are not currently economical to recover, or whose uses are not yet evident) may become more valuable. This assumption leads [the EPA] to the conclusions that it is reasonable to project oil and gas drilling rates (based on the historical record) over the regulatory time frame; and second, that since these rates are surrogates for other potential resources, it is inappropriate to include consequences of activities or secondary recovery techniques specific to oil drilling. (EPA, 1996, pp. 12-7, 12-8)

In effect, the EPA assumes that resources equivalent to more than 100 oil and gas fields, comparable to those now being exploited, will be explored for and developed at the WIPP site by borehole extraction methods, using the drilling density of the past century over the 10,000-year regulatory time frame. (Note: the mining of potash, using techniques in current use near WIPP, would not involve borehole extraction.)

The EPA prescription in 40 CFR 194 with respect to human intrusion illustrates both the difficulty and the subjective judgment involved in arriving at a regulatory standard for WIPP. Prescribing a future frequency of intrusive drilling based on past and present-day technologies is inherently speculative, highly subjective, and untestable by the scientific method (NRC, 1995). It is disconcerting to the committee that a repository such as WIPP, which appears to be an excellent choice based on geological considerations, could be judged "non-compliant" solely on the basis of a criterion which has a poor scientific basis.

The committee considers the approach proposed in the NRC (1995) report, *Technical Bases for Yucca Mountain Standards* (TYMS), to be a more appropriate way in which to assess the significance of human intrusion. The TYMS report recommends that the consequences of a single or combined drilling intrusion (such as E1E2—see Figure 2.5 of this report) be examined qualitatively, rather than quantitatively as in 40 CFR 191, in order to assess the severity of the consequences and to indicate what repository design measures may be appropriate to reduce the severity.

The incremental risk assumed by accepting human intrusion could also be compared with the long-term risks associated with leaving TRU waste on the surface.

RETROSPECTIVE

The WIPP Compliance Certification Application is the first test of a unique legal framework, which necessarily will not be based on prior experience. Inevitably, such a legal framework will contain controversial and weak elements. DOE and EPA will need to work closely, in the public interest, to ensure that such elements of the complex legal framework are resolved expeditiously, through negotiation based upon scientific understanding.

WIPP also is a pioneering effort in the assessment of geological site suitability and design procedures for a waste repository. It is the first repository in the nation for which an application to begin permanent, deep, geological disposal is being submitted for a decision.

The scientific and technical questions posed, particularly to earth scientists, required unprecedented levels of quantitative assessment of the response of a rock mass over tens of thousands to hundreds of thousands of years. A great deal of original research was necessary to demonstrate the applicability of laboratory data to field-scale conditions and to develop and assess the reliability of computer models to reproduce these complex systems. Teams of specialists were assembled, and scientific programs defined. Total System Performance Assessment was introduced and refined. Inevitably perhaps, the scientific emphases, budget allocations, and the most critical information needs identified by PA were not always in concert. These difficulties notwithstanding, a great deal has been accomplished.

In retrospect, one valuable lesson from WIPP is the importance of establishing site-specific PA early in the repository site characterization and design process, both to identify the critical research needs and to establish the overall goals of the project. Scientific programs, priorities, and budget allocations must be identified and periodically re-examined, all in the context of PA.

Significant uncertainties are implicit in extrapolating from one or two decades of data, at most, to tens or hundreds of thousands of years—but a well coordinated integrated program, using PA effectively, should be able to produce sufficiently credible answers to establish, with "reasonable expectation" and within a finite time and budget, whether or not a proposed repository site and design are appropriate for effective long-term isolation of radioactive waste.

References

40 CFR 191. 1995. Environmental radiation protection standards for management and disposal of spent nuclear fuel, high-level and transuranic radioactive wastes. Code of Federal Regulations Title 40, Pt. 191.

40 CFR 194. 1996. Criteria for the certification and recertification of the Waste Isolation Pilot Plant's compliance with the 40 CFR Part 191 disposal regulations; final rule. Federal Register 61(28) (February 9):5224-5245.

40 CFR 268. 1995. Land disposal restrictions. Code of Federal Regulations Title 40, Pt. 268.

Ahlbom, K., J. Andersson, P. Andersson, T. Ittner, C. Ljunggren, and S. Tirén. 1992. Finnsjön Study Site. Scope of Activities and Main Results: SKB Technical Report 92-33. Stockholm: Swedish Nuclear Fuel and Waste Management Company.

Ahrens, E. H., and F. D. Hansen. 1995. Large-Scale Dynamic Compaction Demonstration Using WIPP Salt: Fielding and Preliminary Results. SAND95-1941. Albuquerque, N.Mex.: Sandia National Laboratories.

Anderson, M. P., and W. W. Woessner. 1992. Applied Groundwater Modeling: Simulation of Flow and Advective Transport. San Diego: Academic Press.

Barker, J. M., and G. S. Austin. 1995. Overview of the Carlsbad Potash District. Pp. III-1–III-26 in Evaluation of Mineral Resources at the Waste Isolation Pilot Plant (WIPP) Site, Vol. 2. Socorro, N.Mex.: New Mexico Bureau of Mines and Mineral Resources, Campus Station.

Beauheim, R. L. 1987. Interpretations of Single-Well Hydraulic Tests Conducted at and near the Waste Isolation Pilot Plant (WIPP) Site, 1983-1987. SAND87-0039. Albuquerque, N.Mex.: Sandia National Laboratories.

Beauheim, R. L. 1989. Interpretation of H-11b4 Hydraulic Tests and the H-11 Multipad Pumping Test of the Culebra Dolomite at the Waste Isolation Pilot Plant (WIPP) Site. SAND89-0536. Albuquerque, N.Mex.: Sandia National Laboratories.

Beauheim, R. L., G. J. Saulnier, Jr., and J. Avis. 1991. Interpretation of Brine-Permeability Tests of the Salado Formation at the Waste Isolation Pilot Plant Site: First Interim Report. SAND90-0083. Albuquerque, N.Mex.: Sandia National Laboratories.

Beauheim, R. L., R. M. Roberts, T. F. Dale, M. D. Fort, and W. A. Stensrud. 1993a. Hydraulic Testing of Salado Formation Evaporites at the Waste Isolation Pilot Plant: Second Interpretive Report. SAND92-0533. Albuquerque, N.Mex.: Sandia National Laboratories.

Beauheim, R. L., W. W. Wawersik, and R. M. Roberts. 1993b. Coupled permeability and hydrofracture tests to assess the waste-containment properties of fractured anhydrite. International Journal of Rock Mechanics and Mining Sciences & Geomechanics Abstracts 30, 7: 1159-1163.

Beauheim, R. L., S. M. Howarth, P. Vaughn, S. W. Webb, and K.W. Larson. 1995. Integrating Modeling and Experimental Programs to Predict Brine and Gas Flow at the Waste Isolation Pilot Plant. SAND94-0599A and CONF-941053-1. Albuquerque, N.Mex.: Sandia National Laboratories.

Bein, A., S. D. Hovorka, R. S. Fisher, and E. Roedder. 1991. Fluid inclusions in bedded Permian halite, Palo Duro Basin, Texas: evidence for modification of seawater in evaporite brine-pools and subsequent early diagenesis. Journal of Sedimentary Petrology 61(1):1-14.

Berest, P., and B. Brouard. 1996. Behavior of sealed solution–mined caverns. Paper presented at SALT-IV—Fourth Conference on the Mechanical Behavior of Salt, Montreal, Canada, June 17-18. (Proceedings to be published in early 1997. Clausthal-Zellerfeld, Germany: Trans Tech Publications.)

Borns, D. J. 1995. Implications of geophysical surveys in the WIPP underground on the interpretation of the relative roles of the three proposed conceptual models for Salado fluid flow. Memo Oct. 3, 1994. Appendix I in Systems Prioritization Method—Iteration 2. Baseline Position Paper: Salado Formation Fluid Flow and Transport Contaminant Group. S. Howarth et al., eds. Sandia National Laboratories.

Bottrell, S. H., ed. 1996. Proceedings of the Fourth International Symposium on the Geochemistry of the Earth's Surface. Leeds, U.K.: University of Leeds.

Brady, B. H. G. and E. T. Brown. 1987. Rock Mechanics for Underground Mining. London: Routledge, Chapman and Hall.

Bredehoeft, J. D. 1988. Will salt repositories be dry? EOS (Transactions American Geophysical Union) 69(9):121-131.

Bredehoeft, J. D., and L. F. Konikow. 1993. Reply to comment by G. de Marsily, P. Combes, and P. Goblet. Advances in Water Resources 15(6):371-372.

Brinster, K. F. 1991. Preliminary Geohydrologic, Conceptual Model of the Los Medaños Region Near the Waste Isolation Pilot Plant for the Purpose of Performance Assessment. SAND89-7147. Albuquerque, N.Mex.: Sandia National Laboratories.

Broadhead, R. F., F. Luo, and S. W. Speer. 1995. Oil and Gas Resource Estimates. Pp. XI-1–XI-135 in Evaluation of Mineral Resources at the Waste Isolation Pilot Plant (WIPP) Site, Vol. 3. Socorro, N.Mex.: New Mexico Bureau of Mines and Mineral Resources, Campus Station.

Brodsky, N. S. 1995. Thermodynamical Damage Recovery Parameters for Rocksalt from the Waste Isolation Pilot Plant. SAND93-7111. Albuquerque, N.Mex.: Sandia National Laboratories.

Brodsky, N. S., F. D. Hansen, and T. W. Pfeifle. 1996. Properties of Dynamically Compacted WIPP Salt. SAND96-0838C. Albuquerque, N Mex.: Sandia National Laboratories. Paper presented at SALT IV— Fourth Conference on the Mechanical Behavior of Salt, Montreal, Canada, June 17-18. (Proceedings to be published in early 1997. Clausthal-Zellerfeld, Germany: Trans Tech Publications.)

Brush, L. H. 1990. Test Plan for Laboratory and Modeling Studies of Repository and Radionuclide Chemistry for the Waste Isolation Pilot Plant. SAND90-0266. Albuquerque, N.Mex.: Sandia National Laboratories.

Brush, L. H. 1994. Position Paper on Gas Generation in the Waste Isolation Pilot Plant. Draft for Environmental Protection Agency and Stakeholder Review. Albuquerque, N.Mex.

Brusseau, M. L. 1994. Transport of reactive contaminants in heterogeneous porous media. Reviews of Geophysics 32(3):285-313.

Butcher, B. M. 1990. Preliminary Evaluation of Potential Engineered Modifications for the Waste Isolation Pilot Plant (WIPP). SAND89-3095. Albuquerque, N.Mex.: Sandia National Laboratories.

Butcher, B. M., and F. T. Mendenhall. 1993. A Summary of the Models Used for the Mechanical Response of Disposal Rooms in the Waste Isolation Pilot Plant with Regard to Compliance with 40 CFR 191, Subpart B. SAND92-0427. Albuquerque, N.Mex.: Sandia National Laboratories.

Callahan, G. D., M. C. Loken, L. D. Hurtado, and F. D. Hansen. 1996. Evaluations of constitutive models for crushed salt. Paper presented at SALT-IV—Fourth Conference on the Mechanical Behavior of Salt. June 17-18, Montreal, Canada. (Proceedings to be published in early 1997. Clausthal-Zellerfeld, Germany: Trans Tech Publications.)

Chapman, J. B. 1988. Chemical and Radiochemical Characteristics of Groundwater in the Culebra Dolomite, Southeastern New Mexico. EEG-39. Albuquerque, N.Mex.: Environmental Evaluation Group.

Churchill, R. V. 1972. Operational Mathematics. McGraw-Hill Publishing Company.

Churchill, R. V., and J. W. Brown. 1987. Fourier Series and Boundary Value Problems. McGraw-Hill, Inc.

Cranwell, R. M., R. V. Guzowski, J. E. Campbell, and N. R. Ortiz. 1990. Risk Methodology for Geologic Disposal of Radioactive Waste: Scenario Selection Procedure SAND80-1429. Sandia National Laboratories, prepared for U.S. Nuclear Regulatory Commission, NUREG/CR-1667.

Dale, T., and L. D. Hurtado. 1996. WIPP Air-Intake Shaft Disturbed Rock Zone Study. Paper presented at SALT-IV—Fourth Conference on the Mechanical Behavior of Salt, Montreal, Canada, June 17-18. (Proceedings to be published in early 1997. Clausthal-Zellerfeld, Germany: Trans Tech Publications.)

Davies, P. B. 1989. Variable-Density Ground Water Flow and Paleohydrology in the Waste Isolation Pilot Plant (WIPP) Region, Southeastern New Mexico. U.S. Geological Survey Open-File Report 88-490. Albuquerque, N.Mex.: U. S. Geological Survey.

Davies, P. B., J. F. Pickens, and R. L. Hunter. 1991. Complexity in the Validation of Ground Water Travel Time in Fractured Flow and Transport Systems. SAND89-2379. Albuquerque, N.Mex.: Sandia National Laboratories.

De Marsily, G., P. Combes, and P. Goblet. 1992. Comment on "Ground water models cannot be validated" by L. F. Konikow and J.D. Bredehoeft. Advances in Water Resources 15(6):371-372.

Deal, D. E., R. J. Abitz, D. S. Belski, J. B. Clark, M. E. Crawley, and M. L. Martin. 1991a. Brine Sampling and Evaluation Program 1989 Report. Carlsbad, N.Mex.: Westinghouse Electric Corporation, Waste Isolation Division.

Deal, D. E., R. J. Abitz, J. Myers, J. B. Case, D. S. Belski, M. L. Martin, and W. M. Roggenthem. 1991b. Brine Sampling and Evaluation Program 1990 Report. Carlsbad, N.Mex.: Westinghouse Electric Company, Waste Isolation Division.

Detournay, E., and A. H.-D. Cheng. 1988. Poroelastic response of a borehole in a non-hydrostatic stress field. International Journal of Rock Mechanics and Mining Sciences and Geomechanics Abstracts 25(3):171-182.

DOE (U.S. Department of Energy). 1990. Final Supplement, Environmental Impact Statement, Waste Isolation Pilot Plant. DOE/EIS-0026-FS. Office of Environmental Restoration and Waste Management, Washington, D.C.

DOE. 1991. Evaluation of the Effectiveness and Feasibility of the Waste Isolation Pilot Plant Engineered Alternatives: Final Report of the Engineered Alternatives Task Force, Vol. I and II. DOE/WIPP 91-007, Revision 0. Carlsbad, N.Mex.: WIPP Project Office.

DOE. 1995a. Draft Title 40 CFR 191 Compliance Certification Application for the Waste Isolation Pilot Plant. DRAFT-DOE/CAO-2056. March 31 and update, July 31. Carlsbad, N.Mex.: Carlsbad Area Office.

DOE. 1995b. Engineered Alternatives Cost/Benefit Study Final Report. DOE/WIPP 95-2135, Revision 0. Carlsbad, N.Mex.: Carlsbad Area Office.

DOE. 1995c. Transuranic Waste Baseline Inventory Report. Revision 2. DOE/CAO 95-1121, 3 vols. Carlsbad, N.Mex.: Carlsbad Area Office.

DOE. 1995d. Waste Isolation Pilot Plant Sealing System Design Report. DOE/WIPP-95-3117.

EPA (U.S. Environmental Protection Agency), Office of Radiation Programs. 1988. Limiting Values of Radionuclide Intake and Air Concentration and Dose Conversion Factors for Inhalation, Submersion, and Ingestion. 1988 Federal Guidance Report No. 11, EPA-520/1-88-020. Washington, D.C.

EPA, Office of Radiation and Indoor Air. 1996. Response to Comments. 40 CFR 194: Criteria for the Certification and Re-Certification of the Waste Isolation Pilot Plant's Compliance with the 40 CFR Part 191 Disposal Regulations. EPA 402-R-96-001.

Eyermann, T. J., L. L. Van Sambeek, and F. D. Hansen. 1995. Case Studies of Sealing Methods and Material Used in the Salt and Potash Mining Industries. SAND95-1120. Albuquerque, N.Mex.: Sandia National Laboratories.

Felmy A. R., and J. H. Weare. 1986. The prediction of borate mineral equilibria in natural waters: application to Searles Lake, California. Geochimica et Cosmochimica Acta 50:2771-2783.

Finley, R. E., and J. R. Tillerson. 1992. WIPP Small Scale Seal Performance Tests—Status and Impacts. SAND91-2247. Albuquerque, N.Mex.: Sandia National Laboratories.

Franke, O. L., T. E. Reilly, and G. D. Bennett. 1987. Definition of boundary and initial conditions in the analysis of saturated ground water flow systems—An introduction. Ch. B5 in Techniques of Water-Resources Investigations of the United States Geological Survey, Book 3: Applications of Hydraulics. U.S. Geological Survey.

Freeze, G. A., and T. L. Christian-Frear. In preparation. Modeling Brine Inflow to Room Q. A Numerical Investigation of Flow Mechanics.

Freeze, G. A., K. W. Larson, and P. B. Davies. October 1995a. Coupled Multiphase Flow and Closure Analysis of Repository Response to Waste-Generated Gas at the Waste Isolation Pilot Plant (WIPP). SAND93-1986. Sandia National Laboratories. Albuquerque, N.Mex.

Freeze, G. A., K. W. Larson, and P. B. Davies. 1995b. A Summary of Methods for Approximating Salt Creep and Disposal Room Closure in Numerical Models of Multiphase Flow. SAND94-0251. Albuquerque, N.Mex.: Sandia National Laboratories.

Garabedian, S. P., D. R. LeBlanc, L. W. Gelhar, and M. A. Celia. 1991. Large-scale natural gradient tracer test in sand and gravel, Cape Cod, Massachusetts: 2, analysis of tracer moments for a nonreactive tracer. Water Resources Research 27(5):911-924.

Garven, G. 1995. Continental-scale groundwater flow and geologic processes. Annual Review of Earth and Planetary Sciences 23:89-117.

Gelhar, L. W., C. Welty, and K. R. Rehfeldt. 1992. A critical review of data on field-scale dispersion in aquifers. Water Resources Research 28(7):1955-1974.

Goode, D. J., and L. F. Konikow. 1989. Modification of a Method-of-Characteristics Solute-Transport Model to Incorporate Decay and Equilibrium-Controlled Sorption or Ion Exchange. U.S. Geological Survey Water-Resources Investigations Report 89-4030.

Grenthe, I., and H. Wanner. 1992. Guidelines for the Extrapolation to Zero Ionic Strength. NEA-TBD-2, revision 2. Gif-sur-Yvette, France: Nuclear Energy Agency, Organization for Economic Co-operation and Development.

Griswold, G. B. 1995. Future Mining Technology Pp. IV-1–IV-5 in Evaluation of Mineral Resources at the Waste Isolation Pilot Plant (WIPP) Site. Vol. 2. Socorro, N.Mex.: New Mexico Bureau of Mines and Mineral Resources, Campus Station.

Gulick, C. W., Jr., J. A. Boa, Jr., D. M. Walley, and A. D. Buck. 1980. Borehole Plugging Materials Development Program, Report 2. SAND79-1514. Albuquerque, N.Mex.: Sandia National Laboratories.

Hansen, F. D., and E. H. Ahrens. 1996. Large-Scale Dynamic Compaction of Natural Salt. SAND96-0792C. Albuquerque, N.Mex.: Sandia National Laboratories. Paper presented at SALT IV—Fourth Conference on the Mechanical Behavior of Salt, Montreal, Canada, June 17-18. (Proceedings to be published in early 1997. Clausthal-Zellerfeld, Germany: Trans Tech Publications.)

Hareland, G. 1995. The Past-Decade Developments and Future Trends in Oil-Well Drilling, Completion, and Stimulation, with Special Applications to Developments at the WIPP Site. Chapter X in Evaluation of Mineral Resources at the Waste Isolation Pilot Plant (WIPP) Site, Vol. 3. Socorro, N.Mex.: New Mexico Bureau of Mines and Mineral Resources, Campus Station.

Harvie C. E., N. Møller, and J. H. Weare. 1984. The Prediction of Mineral Solubilities in Natural Waters: The $Na-K-Mg-Ca-H-Cl-SO_4-OH-HCO_3-CO_3-CO_2-H_2O$ system to high ionic strengths at 25°C. Geochimica et Cosmochimica Acta 48:723-751.

Helton, J. C., J. W. Garner, R. D. McCurley, and D. K. Rudeen. 1991. Sensitivity Analysis Techniques and Results for Performance Assessment at the Waste Isolation Pilot Plant. SAND90-7103. Albuquerque, N.Mex.: Sandia National Laboratories.

Howarth, S., K. Larson, T. Christian-Frear, R. Beauheim, D. Borns, D. Deal, A. L. Jensen, K. Knowles, D. Powers, R. Roberts, M. Tierney, and S. Webb. 1995. Systems Prioritization Method—Iteration 2. Baseline

Position Paper: Salado Formation Fluid Flow and Transport Contaminant Group. Albuquerque, N.Mex.: Sandia National Laboratories.

Ibaraki, M., and E. A. Sudicky. 1995. Colloid-facilitated contaminant transport in discretely fractured porous media. 1. Numerical formulation and sensitivity analysis. Water Resources Research 31: 2945-2960.

Jensen, A. L., R. L. Jones, E. N. Lorusso, and C. L. Howard. 1993. Large-Scale Brine Inflow Data Report for Room Q Prior to November 25, 1991. SAND92-1173. Albuquerque, N.Mex.: Sandia National Laboratories.

Jones, B. F., and S. K. Anderholm. 1993. Normative analysis of brines from the Salado Formation and underlying strata, SE New Mexico. EOS (Transactions of the American Geophysical Union) 74(16):326. Abstract.

Jones, B. F., and S. K. Anderholm. 1996. Some geochemical considerations of brines associated with bedded salt repositories. Pp. 343-353 in Proceedings of the Fourth International Symposium on the Geochemistry of the Earth's Surface, S. H. Bottrell, ed. Leeds, U.K.: University of Leeds.

Kaplan, S., and B. J. Garrick. 1981. On the quantitative definition of risk. Risk Analysis 1(1):11-27.

Kay, M., and E. H. Colbert. 1965. Stratigraphy and Life History. New York: John Wiley and Sons.

Kelley, V. A., and G. J. Saulnier, Jr. 1990. Core Analyses for Selected Samples from the Culebra Dolomite at the Waste Isolation Pilot Plant Site. SAND90-7011. Albuquerque, N.Mex.: Sandia National Laboratories.

Kim, J. I. 1994. Actinide colloids in natural aquifer systems. Materials Research Society Bulletin XIX(12):47-53.

Konikow, L. F., and J. D. Bredehoeft. 1992. Ground water models cannot be validated. Advances in Water Resources 15(1):75-83.

Kushner, D. J. 1978. Life in High Salt and Solute Concentrations—Halophilic Bacteria. Pp. 317-335 in Microbial Life in Extreme Environments, D. J. Kushner, ed. New York: Academic Press.

Laaksoharju, M., C. Degueldre, and C. Skårman. 1995. Studies of Colloids and Their Importance for Repository Performance Assessment. SKB Technical Report 95-24. Stockholm: Swedish Nuclear Fuel and Waste Management Company.

Lappin, A. R. 1988. Summary of Site-Characterization Studies Conducted from 1983 Through 1987 at the Waste Isolation Pilot Plant (WIPP) Site, Southeastern New Mexico. SAND88-0157. Albuquerque, N.Mex.: Sandia National Laboratories.

Lappin, A. R., and R. L. Hunter, eds. 1989. Systems Analysis, Long-Term Radionuclide Transport, and Dose Assessments, Waste Isolation Pilot Plant (WIPP), Southeastern New Mexico. SAND89-0462. Albuquerque, N.Mex.: Sandia National Laboratories.

LaVenue, A. M., T. L. Cauffman, and J. F. Pickens. 1990. Ground Water Flow Modeling of the Culebra Dolomite, Vol. 1: Model Calibration. SAND89-7068/1. Albuquerque, N.Mex.: Sandia National Laboratories.

LaVenue, A. M., B. S. RamaRao, G. de Marsily, and M. G. Marietta. 1995. Pilot point methodology for automated calibration of an ensemble of conditionally simulated transmissivity fields, 2. Application. Water Resources Research 31(3):495-516.

Lienert, C., S. A. Short, and H. R. von Gunten. 1994. Uranium Infiltration from a River to Shallow Groundwater. Geochimica et Cosmochimica Acta 58(24):5455-5463.

Marietta, M. C., S. G. Bertram-Howery, D. R. Anderson, K. F. Brinster, R. V. Guzowski, H. Lussolino, and R. P. Rechard. 1989. Performance Assessment Methodology Demonstration: Methodology Development for Evaluation Compliance with EPA 40 CFR 191, Subpart B, for the Waste Isolation Pilot Plant. SAND89-2027. Albuquerque, N.Mex.: Sandia National Laboratories.

Marshal, K. 1976. Interfaces in Microbial Ecology. Cambridge, Mass.: Harvard University Press.

McKinley, I. G., and D. Savage. 1993. Comparison of Solubility Databases Used for HLW Performance Assessment. Radiochimica Acta 66/67:657-665.

McTigue, D. F. 1991. Horizontal Darcy Flow to Room Q. Memo June 24. Appendix E in Systems Prioritization Method—Iteration 2. Baseline Position Paper: Salado Formation Fluid Flow and Transport Contaminant Group, S. Howarth et al., eds. Albuquerque, N.Mex.: Sandia National Laboratories.

McTigue, D. F. 1993. Permeability and Hydraulic Diffusivity of Waste Isolation Pilot Plant Repository Salt Inferred from Small-Scale Brine Inflow Experiments. SAND92-1911. Albuquerque, N.Mex.: Sandia National Laboratories.

McTigue, D. F. 1995a. Calculation of brine flux and cumulative brine volume for room Q, based on a Darcy-flow model. Memo April 3, 1989. Pp. E-35–E-36 in Systems Prioritization Method—Iteration 2. Baseline Position Paper: Salado Formation Fluid Flow and Transport Contaminant Group, S. Howarth et al., eds. Albuquerque, N.Mex.: Sandia National Laboratories.

McTigue, D. F. 1995b. A model for brine inflow due to salt "damage." Memo August 2, 1990. Pp. E-3–E-29 in Systems Prioritization Method Iteration 2. Baseline Position Paper: Salado Formation Fluid Flow and Transport Contaminant Group, S. Howarth et al., eds. Albuquerque, N.Mex.: Sandia National Laboratories.

Mercer, J. W. 1987. Compilation of Hydrologic Data from Drilling the Salado and Castile Formations near the Waste Isolation Pilot Plant (WIPP) Site in Southeastern New Mexico. SAND86-0954. Albuquerque, N.Mex.: Sandia National Laboratories.

Meyer, D., and J. J. Howard, eds. 1983. Evaluation of Clays and Clay Minerals for Application to Repository Sealing. Prepared for the Office of Nuclear Waste Isolation (ONWI), Columbus, Ohio, by D'Appolonia and Pennsylvania State University.

Miller, C. W., and L. V. Benson. 1983. Simulation of solute transport in a chemically reactive heterogeneous system: model development and application. Water Resources Research 19(2):381-391.

Moench, A. F. 1984. Double-porosity models for a fissured groundwater reservoir with fracture skin. Water Resources Research (20):834-846.

Molecke, M. A. 1979a. Gas Generation from Transuranic Waste Degradation: Data Summary and Interpretation. SAND79-1245. Sandia National Laboratories, Albuquerque, N.Mex.

Molecke, M. A. 1979b. Gas generation potential from TRU wastes. Pp. 3-1–3-21 in Summary of Research and Development Activities in Support of Waste Acceptance Criteria for WIPP. SAND79-1305. Sandia National Laboratories. Albuquerque, N.Mex.

Munson, D. E. In press. Constitutive Model of Creep in Rock Salt Applied to Underground Room Closure. International Journal of Rock Mechanics and Mining Sciences and Geomechanics Abstracts.

Munson, D. E., D. J. Borns, M. K. Pickens, D. J. Holcomb, and S. E. Bigger. 1995. Systems Prioritization Method—Iteration 2 Baseline Position Paper: Rock Mechanics: Creep, Fracture, and Disturbed Rock Zone (DRZ). Albuquerque, N.Mex.: Sandia National Laboratories.

Neill, R. H., Lokesh C., W. W.-L. Lee, T. M. Clemo, M. K. Silva, J. W. Kenney, W. T. Bartlett, B. A. Walker. 1996. Review of the WIPP Draft Application to Show Compliance with EPA Transuranic Waste Disposal Standards. EEG-61. Environmental Evaluation Group, N.Mex.

Nichols, M. D., and E. P. Laws. 1995. Letters to Senators Larry E. Craig and Dirk Kempthorne. U.S. Environmental Protection Agency.

Nitsche, H., K. Roberts, R. Xi, T. Prussin, K. Becraft, I. Al Mahamid, H. B. Silver, S. A. Carpenter, R. C. Gatti, and C. F. Novak. 1994. Long term plutonium solubility and speciation studies in a synthetic brine. Radiochimica Acta 66/67: 3-8.

Novak, C. F. 1992. An Evaluation of Radionuclide Batch Sorption Data on Culebra Dolomite for Aqueous Compositions Relevant to the Human Intrusion Scenario for the Waste Isolation Pilot Plant. SAND91-1299. Albuquerque, N.Mex.: Sandia National Laboratories.

Novak, C. F. 1995a. Actinide Chemistry Research Supporting the Waste Isolation Pilot Plant (WIPP): FY94 Results. SAND94-2274. Albuquerque, N.Mex.: Sandia National Laboratories.

Novak, C. F. 1995b. The Waste Isolation Pilot Plant (WIPP) Actinide Source Term: Test Plan for the Conceptual Model and the Dissolved Concentration Submodel. SAND95-1985. Albuquerque, N.Mex.: Sandia National Laboratories.

Novak, C. F., and K. E. Roberts. 1995. Thermodynamic Modeling of Neptunium (V) Solubility in Na-CO_3-HCO_3-Cl-ClO_4-H-OH-H_2O Electrolytes. Scientific Basis for Nuclear Waste Management XVIII. Proceedings of the Materials Research Society, 353:1119-1128.

Novak, C. F., R. F. Weimer, H. W. Papenguth, Y. K. Behl, D. A Lucero, F. Gelbard, and J. A. Romero. Unpublished. Actinide Source Term and Chemical Retardation Programs letter and briefing materials on solubility, colloid, and chemical retardation work, distributed at Actinide Source Term and Chemical Retardation Programs meeting. June 29, 1995, Albuquerque, N.Mex.

NRC (National Research Council). 1957. The Disposal of Radioactive Waste on Land. Washington, D.C.: National Academy Press.

NRC. 1970. Disposal of Solid Radioactive Wastes in Bedded Salt Deposits. Washington, D.C.: National Academy Press.

NRC. 1979a. WIPP Letter Report to Mr. Sheldon Meyers, Program Director of the U.S. DOE Office of Nuclear Waste Management. Washington, D.C.: National Academy Press.

NRC. 1979b. WIPP Letter Report to Mr. Sheldon Meyers, Program Director of the U.S. DOE Office of Nuclear Waste Management. Washington, D.C.: National Academy Press.

NRC. 1983. A Study of the Isolation System for Geologic Disposal of Radioactive Waste. Washington, D.C.: National Academy Press.

NRC. 1984. Review of the Scientific and Technical Criteria for the Waste Isolation Pilot Plant. Washington, D.C.: National Academy Press.

NRC. 1987. WIPP Letter Report to Mr. John Mathur, U.S. DOE Office of Defense Waste and Transportation Management. Washington, D.C.: National Academy Press.

NRC. 1988a. WIPP Letter Report to the Honorable John S. Herrington, Secretary of the Department of Energy. Washington, D.C.: National Academy Press.

NRC. 1988b. WIPP Letter Report to Mr. Critz George, U.S. DOE Office of Defense Waste and Transportation Management. Washington, D.C.: National Academy Press.

NRC. 1989. WIPP Letter Report. Washington, D.C.: National Academy Press.

NRC. 1991. WIPP Letter Report to Mr. Leo Duffy, Director of the U.S. DOE Office of Environmental Restoration and Waste Management. Washington, D.C.: National Academy Press.

NRC. 1992. WIPP Letter Report (addressing the underground experimental plan with TRU wastes) to the Honorable Leo P. Duffy, Assistant Secretary of the U.S. DOE Office of Environmental Restoration and Waste Management. Washington, D.C.: National Academy Press.

NRC. 1995. Technical Bases for Yucca Mountain Standards. Washington, D.C.: National Academy Press.

Papenguth, H. W. and Y. K. Behl. 1996a. Test Plan for Evaluation of Colloid-Facilitated Actinide Transport at the WIPP. Test Plan 96-01. Albuquerque, N.Mex.: Sandia National Laboratories.

Papenguth, H. W., and Y. K. Behl. 1996b. Test Plan for Evaluation of Dissolved Actinide Retardation at the WIPP. Test Plan 96-02. Albuquerque, N.Mex.: Sandia National Laboratories.

Pickard, Lowe and Garrick, Inc.; Westinghouse Electric Corporation; and Fauske and Associates, Inc. 1981. Zion Probabilistic Safety Study. Prepared for Commonwealth Edison Company.

Pitzer, K. S. 1973. Thermodynamics of electrolytes. I. Theoretical basis and general equations. J. Phys. Chem. 77: 268-277.

Pitzer, K. S. 1977. Electrolyte theory—Improvements Since Debye and Hückel. Acct. Chem. Res., 10: 371-377.

Pitzer, K. S., ed. 1991. Activity Coefficients in Electrolyte Solutions. 2nd ed. Boca Raton, Fla.: CRC Press.

Powers, D. W., S. J. Lambert, S.-E. Shaffer, L. R. Hill, and W. D. Weart, eds. 1978. Geological Characterization Report, Waste Isolation Pilot Plant (WIPP) Site, Southeastern New Mexico, Vols. I and II. SAND78-1596. Albuquerque, N.Mex.: Sandia National Laboratories.

Prindle, N. H., F. T. Mendenhall, D. M. Boak, W. Beyeler, D. Rudeen, R. C. Lincoln, K. Trauth, D. R. Anderson, M. G. Marietta, and J. C. Helton. 1996. The Second Iteration of the Systems Prioritization Method: A Systems Prioritization and Decision-Aiding Tool for the Waste Isolation Pilot Plant. 3 vols. SAND95-2017. Albuquerque, N.Mex.: Sandia National Laboratories.

Reardon, E. J. 1981. Kd's—Can they be used to describe reversible ion sorption reactions in contaminant migration? Ground Water 19(3):279-286.

Rechard, R. P. 1995. An Introduction to the Mechanics of Performance Assessment Using Examples of Calculations Done for the Waste Isolation Pilot Plant Between 1990 and 1992. SAND93-1378. Albuquerque, N. Mex.: Sandia National Laboratories.

Reed, D. T., S. Okajima, and M. K. Richmann. 1994. Stability of Plutonium(VI) in Selected WIPP Brines. Radiochimica Acta 66/67:95-101.

Reilly, T. E., O. L. Franke, H. T. Buxton, and G. D. Bennett. 1987. A Conceptual Framework for Ground Water Solute-Transport Studies with Emphasis on Physical Mechanisms of Solute Movement. U.S. Geological Survey Water-Resources Investigation Report 87-4191.

RE/SPEC. Unpublished. Calculations done under contract to Sandia National Laboratories WIPP Shaft Sealing Program. RSI Calculation File 325/11/03.

RE/SPEC. Unpublished. Calculations done under contract to Sandia National Laboratories WIPP Shaft Sealing Program. RSI Calculation File 325/11/04.

RE/SPEC. Unpublished. Calculations done under contract to Sandia National Laboratories WIPP Shaft Sealing Program. RSI Calculation File 325/11/05.

RE/SPEC. 1995. External Memorandum to L. Diane Hurtado, Sandia National Laboratories, November 10. RSI(RCO)-325/11/95/23.

Roddy, J. W., H. C. Claiborne, R. C. Ashline, P. J. Johson, and B. T. Rhyne. 1986. Physical and Decay Characteristics of Commercial LWR Spent Fuel. ORNL/TM-9591/V16R1. Oak Ridge, Tenn.: Oak Ridge National Laboratory.

Roedder, E., W. M. d'Angelo, A. F. Dorrzapf, Jr., and P. J. Aruscavage. 1987. Composition of Fluid Inclusions in Permian Salt Beds, Palo Duro Basin, Texas, U.S.A. Chemical Geology 61:79-90.

Rossmanith, H.-P., ed. 1995. Mechanics of Jointed and Faulted Rock: Proceedings of the 2nd International Conference on the Mechanics of Jointed and Faulted Rock, MJFR-2, Vienna, Austria, 10-14 April, 1995. Rotterdam: A. A. Balkema.

Roy, D. M., M. W. Grutzeck, and L. D. Wakeley. 1983. Selection and Durability of Seal Materials for a Bedded Salt Repository, Preliminary Studies. ONW1-479. Prepared for Office of Nuclear Waste Isolation, Battelle Memorial Institute, Columbus, Ohio, by Pennsylvania State University.

Roy, D. M., M. W. Grutzeck, and L. D. Wakeley. 1985. Salt Repository Seal Materials. A Synopsis of Early Cementitious Materials Development, BMI/ONWI-536. Office of Nuclear Waste Isolation, Battelle Memorial Institute, Columbus, Ohio, by Pennsylvania State University.

Sandia National Laboratories. 1979. Summary of Research and Development Activities in Support of Waste Acceptance Criteria for WIPP. SAND79-1305. Albuquerque, N.Mex.

Sandia National Laboratories. 1991. Preliminary Comparison with 40 CFR Part 191. Subpart B for the Waste Isolation Pilot Plant. December 1991, Vol. 3: Reference Data. SAND91-0893/3. Albuquerque, N.Mex.

Sandia National Laboratories. 1992. Preliminary Performance Assessment for the Waste Isolation Pilot Plant, December 1992, Vols. 1-5. SAND92-0700. Albuquerque, N.Mex.

Sandia National Laboratories. 1995. SPM-2 Report, Revision 1, Vols. I and II. Albuquerque, N.Mex.

Sewards, T., R. Glenn, and K. Keil. 1991. Mineralogy of the Rustler Formation in the WIPP-19 Core. SAND87-7036. Albuquerque, N.Mex.: Sandia National Laboratories.

Sexton, T. J. 1996. Memo, April 11, to Jan van Schilfgaarde, Agricultural Research Service, U.S. Department of Agriculture.

Sharp, J. M., Jr., N. I. Robinson, R. C. Smyth-Boulton, and K. L. Milliken. 1995. Fracture Skin Effects in Groundwater Transport. Pp. 449-454 in Mechanics of Jointed and Faulted Rock, H.-P. Rossmanith, ed. Rotterdam: A. A. Balkema.

Shoesmith, D. W., S. Sunder, M. G. Bailey, and N. H. Miller. 1996. Corrosion of used nuclear fuel in aqueous perchlorate and carbonate solutions. Journal of Nuclear Materials 227:287-299.

Siegel, M. D., and S. Anderholm. 1994. Geochemical evolution of groundwater in the Culebra Dolomite near the Waste Isolation Pilot Plant, southeastern New Mexico, USA. Geochimica et Cosmochimica Acta 58(10):2299-2323.

Siegel, M. D., S. J. Lambert, and K. L. Robinson, eds. 1991. Hydrogeochemical Studies of the Rustler Formation and Related Rocks in the Waste Isolation Pilot Plant Area, Southeastern New Mexico. SAND88-0196. Albuquerque, N.Mex.: Sandia National Laboratories.

Silva, M. K. 1994. Implications of the Presence of Petroleum Resources on the Integrity of the WIPP. EEG-55; DOE/AL/58309-55. Albuquerque, N.Mex.: Environmental Evaluation Group.

Silva, M. K. 1996. Fluid Injection for Salt Water Disposal and Enhanced Oil Recovery as a Potential Problem for the WIPP: Proceedings of a June 1995 Workshop and Analysis. EEG-62. Albuquerque, N.Mex.: Environmental Evaluation Group.

Stormont, J. C. 1984. Plugging and Sealing Program for the Waste Isolation Pilot Plant (WIPP). SAND84-1057. Albuquerque, N.Mex.: Sandia National Laboratories.

Stormont, J. C., ed. 1986. Development and Implementation: Test Series A of the Small-Scale Seal Performance Tests. SAND85-2602. Albuquerque, N.Mex.: Sandia National Laboratories.

Stormont, J. C. 1988. Preliminary Seal Design Evaluation for the Waste Isolation Pilot Plant. SAND87-3083. Albuquerque, N.Mex.: Sandia National Laboratories.

Stormont, J. C., C. L. Howard, and J. J. K. Daemen. 1991. In Situ Measurements of Rock Salt Permeability Changes Due to a Nearby Excavation. SAND90-3134. Albuquerque, N.Mex.: Sandia National Laboratories.

Swedish Nuclear Fuel and Waste Management Company. 1995. Feasibility Study for Siting of a Deep Repository Within the Storuman Municipality: SKB Technical Report 95-08, Stockholm.

Swedish Nuclear Power Inspectorate. 1987. The International HYDROCOIN Project—Background and Results. Paris, France: Organization for Economic Co-operation and Development.

Swedish Nuclear Power Inspectorate. 1990. The International INTRAVAL Project—Background and Results. Paris, France: Organization for Economic Co-operation and Development.

Tanji, K. K., and B. Yaron, eds. 1994. Management of Water Use in Agriculture. New York: Springer-Verlag.

U.S. Congress. 1980. The U.S. Department of Energy National Security and Military Applications of Nuclear Energy Authorization Act of 1980. Public Law 96-164. 93 Stat. 1259. Enacted December 1979. 96th Congress.

U.S. Congress. 1992. Waste Isolation Pilot Plant Land Withdrawal Act P. L. 102-579. Legislative Report for the 102nd Congress.

U.S. Congress, House of Representatives. 1995. Waste Isolation Pilot Plant Land Withdrawal Amendment Act. H.R. 1663. Introduced May 17, 1995, by Representative Skeen (R-N.Mex.). 104th Congress, 1st Session.

U.S. Congress, Senate. 1995. Waste Isolation Pilot Plant Land Withdrawal Amendment Act. S1402. 104th Congress, 1st Session.

U.S. Congress, Senate. 1996. Waste Isolation Pilot Plant Land Withdrawal Amendment Act. Amendment No. 4085 to S1745. 104th Congress. C. R. pp. S6587-S6591.

U.S. Geological Survey. 1987. Techniques of Water-Resources Investigations of the United States Geological Survey, Book 3: Application of Hydraulics.

U.S. Nuclear Regulatory Commission. 1975. Reactor Safety Study: An Assessment of Accident Risks in U.S. Commercial Nuclear Power Plants. WASH-1400, NUREG-75/014.

Valocchi, A. J. 1984. Describing the transport of ion-exchanging contaminants using an effective K_d approach. Water Resources Research 20(4):499-503.

Van der Leeden, F., L. A. Cerrillo, and D. W. Miller. 1975. Ground water pollution problems in the northwestern United States. EPA-660/3-75-018. U.S. Environmental Protection Agency.

Van Sambeek, L. L., D. D. Luo, M. S. Lin, W. Ostrowski, and D. Oyenuga. 1993. Seal Design Alternatives Study. SAND92-7340. Albuquerque, N.Mex.: Sandia National Laboratories.

Wakeley, L. D., P. T. Harrington, and C. A. Weiss, Jr. 1993. Properties of Salt-Saturated Concrete and Grout after Six Years In Situ at the Waste Isolation Pilot Plant. SAND93-7019. Albuquerque, N.Mex.: Sandia National Laboratories.

Wakeley, L. D., T. S. Poole, and J. P. Burkes. 1994. Durability of Concrete Materials in High Magnesium Brine. SAND93-7073. Albuquerque, N.Mex.: Sandia National Laboratories.

Wakeley, L. D., P. T. Harrington, and F. D. Hansen. 1995. Variability in Properties of Salado Mass Concrete. SAND94-1495. Albuquerque, N.Mex.: Sandia National Laboratories.

Waste Isolation Plant Land Withdrawal Amendments Act. 1996. National Defense Authorization Act for Fiscal Year 1997, Subtitle F. (P.L. 104-201, September 23).

Webb, S. W., and K. W. Larson. 1996. The Effect of Stratigraphic Dip on Brine Inflow and Gas Migration at the Waste Isolation Pilot Plant. SAND 94-0932. Albuquerque, N.Mex.: Sandia National Laboratories.

Wikberg, P., G. Gustafson, I. Rhén, and R. Stanfors. 1991. Äspö Hard Rock Laboratory. Evaluation and Conceptual Modeling Based on the Pre-investigations 1986-1990. SKB Technical Report 91-22. Stockholm: Swedish Nuclear Fuel and Waste Management Company.

Williams, N. 1995. The trials and tribulations of cracking the prehistoric code. Science 269:923-924.

Yaron, B., and H. Frenkel. 1994. Water Suitability for Agriculture. In Management of Water Use in Agriculture, K. K. Tanji and B. Yaron, eds. New York: Springer-Verlag.

1. Native yucca in bloom near Carlsbad, New Mexico; the U.S. Department of Energy's Waste Isolation Pilot Plant (WIPP) is in the background.

2. Aerial view of WIPP.

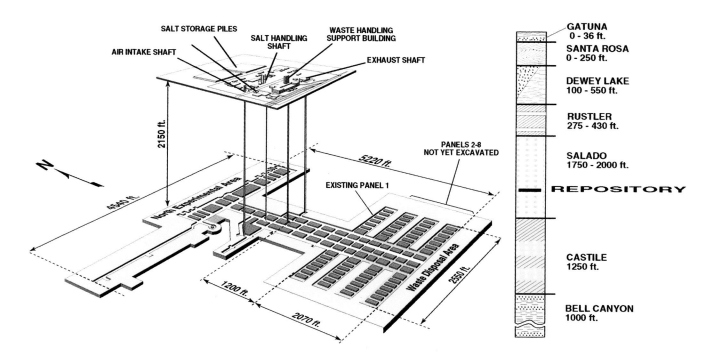

3. Three-dimensional view of the WIPP facility. Panel 1 is the only one of the eight panels shown that has been excavated in the waste disposal area to date. The typical stratigraphy is shown in cross-section on the right of the diagram. WIPP is located in the bedded salt Salado Formation, approximately 660 m below the surface. The Salado is approximately 230 million years old.

4. View of Room 1, Panel 1, underground at WIPP.

5. *A tractor-trailer travels to WIPP with U.S. Nuclear Regulatory Commission-certified containers, using the Transuranic Package Transporter (TRUPACT-II) transportation system. The tractor-trailer and containers are designed specially for shipment of radioactive waste for disposal at WIPP. The location of the TRUPACT-II can be pinpointed within 150 m at any time by a satellite tracking system.*

6. *Stacking of simulated contact-handled transuranic waste approximately 660 m underground at WIPP.*

7. *Conceptual release pathways for the E1E2 human intrusion event (see Chapter 2). It is assumed that at some time in the next 10,000 years, a borehole penetrates both the waste and a region of pressurized brine in the underlying Castile formation. A second borehole is also assumed to penetrate the waste but not the pressurized brine. Arrows indicate the hypothetical direction of ground water flow and radionuclide transport; R_c indicates material released at the surface directly from the drilling operation; R_{acc} indicates radionuclides released below the surface into a permeable bed of Culebra dolomite. This and other release pathways are considered in calculations made to evaluate whether or not a WIPP repository would comply with U.S. Environmental Protection Agency (EPA) standards.*

Source: The figure is taken from the DRAFT-DOE/CAO-2056, p. 6-51 (July 1995 update).

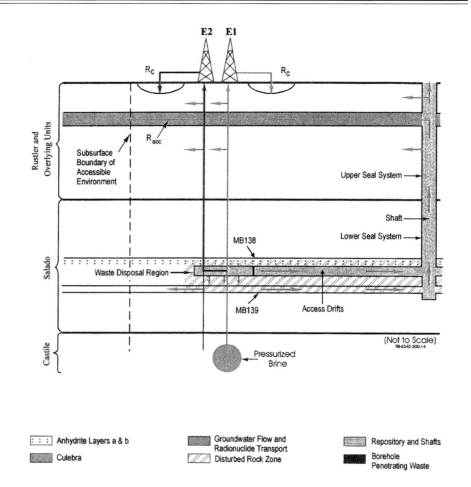

8. *Mean Complementary Cumulative Distribution Functions (CCDFs) showing the probability of radionuclide releases from WIPP over 10,000 years for undisturbed conditions (red) and intrusive drilling scenarios (blue). This figure illustrates the dominant influence of intrusion scenarios on the predicted releases. These preliminary CCDFs are based on early models and data and do not address all the requirements of EPA regulations 40 CFR 191 or 40 CFR 194. The EPA release limits are shown as the left-hand-side limits of the region in green. Compliance is achieved if the mean CCDF never enters the green region.*

Source: The figure is taken from DRAFT-DOE/CAO-2056, p. 6-59 (Vol. I, March 1995) and p. 6-101 (July 1995 update).

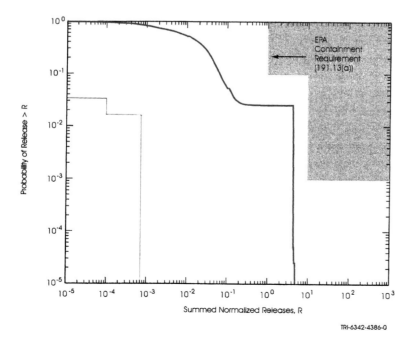

Appendixes

Appendix A

Natural Setting and Resources

As for any deep geologic facility, the natural setting of the Waste Isolation Pilot Plant (WIPP) is important in determining the facility's performance as a repository for radioactive waste. The natural setting consists of the geology, hydrology, geochemistry, climate, and natural resources of the region and local area immediately surrounding WIPP, some of which are discussed below.

GEOLOGIC FRAMEWORK

For a large part of the early geologic history of the North American continent, the land now occupied by southeastern New Mexico was part of an ancient ocean south of the continent. For hundreds of millions of years, this region experienced almost uninterrupted deposition of marine shoreline sediments, primarily beach sands, and shallow water lime and clay muds. In the Pennsylvanian Period and later, coarser continental shelf and slope deposits occasionally slid into deeper basins. The relatively warm, calm seas promoted an abundance of marine life, which, upon death, accumulated in the muds and sands, altered to simple hydrocarbons, and eventually became the oil and gas found in some of these rock layers in recent years.

In pre-Permian periods, the collision of tectonic plates of the earth's mobile crust caused mountains to form to the southeast, north, and west of the WIPP site. The southeastern corner of New Mexico and western Texas remained relatively stable through the Permian Period of earth history (286 million to 245 million years ago [Ma])[1], although some instability caused the earth's crust beneath this region to rise in some places and sink in others, in a belated response to the collision pressures. These fluctuations resulted in shallow areas on the sea floor ("shelves" or "platforms") and deeper areas ("basins") at the beginning of the Permian Period in southeastern New Mexico and western Texas (Figure A.1).

It was in the sediments accumulating in these deep basins that oil was generated, which moved into adjacent shelf environments to form the richest oil pools. One of these deep basins, the Delaware Basin, underlies WIPP and part of southeastern New Mexico. Most of these older rocks beneath WIPP are far too deep to bear on the performance of the repository. However, the presence of oil is considered a possible reason for inadvertent, or even deliberate, human intrusion into the repository in the future.

Late in the Permian, the warm and shallow seas encouraged the formation of reefs, similar to contemporary coral reefs in shallow tropical marine environments. As they grew, the reefs blocked off parts of the sea from time to time. With circulation of seawater restricted by the reef, the seawater began to evaporate, resulting in highly concentrated brines. As the brines continued to evaporate because of the warm temperature and restricted inflow of seawater, crystalline salts began to precipitate and accumulate on the bottom of the restricted basin. The salts varied in chemical composition, depending on the amount of water that had evaporated and on the concentrations of calcium (Ca), magnesium (Mg), sodium (Na), potassium (K), and other components of the common "evaporite" deposits.

As a result of these conditions, the WIPP area is underlain by a total accumulation of hundreds to thousands of meters of reef limestone (calcium carbonate, $CaCO_3$); dolomite [calcium magnesium carbonate—$CaMg(CO_3)_2$]; gypsum [a hydrous (water-bearing) calcium sulfate, $CaSO_4 \cdot 2H_2O$); halite (rock salt, NaCl, the source of common table salt); smaller amounts of anhydrite ($CaSO_4$); and potassium salts (collectively known as "potash," a commercially useful deposit).

---

[1]Older geologic dating extended the Permian to 225 Ma; more recent methods date the end as 245 Ma.

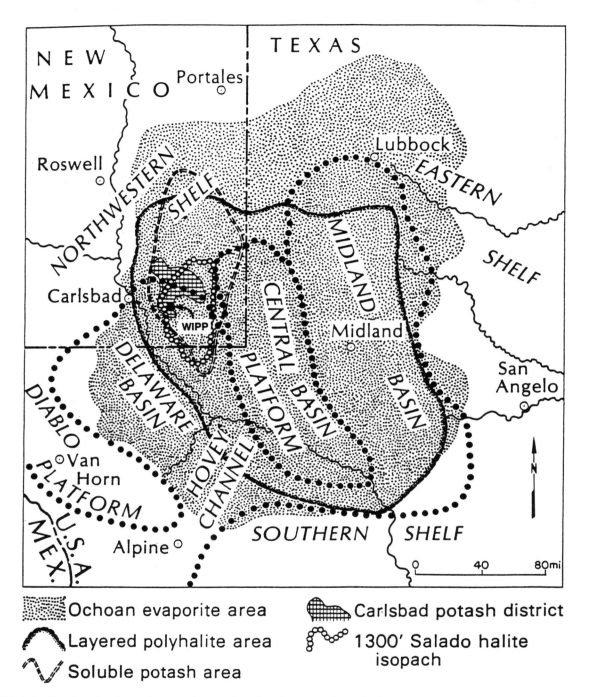

FIGURE A.1 Regional subsurface geology, showing features from before and during the Permian Period. Heavy black dots outline the Pennsylvanian to middle Permian (Guadalupian) shoreline, showing the shallow platforms and shelf areas and the adjoining deeper basins. Superimposed on that background in fine, close-spaced dotted shading is the extent of the late Permian (Ochoan) evaporating sea. A solid black line delineates the extent of layered varied salt deposits. The enclosed form of open circles shows where the Salado Formation salt beds are 396 m (1,300) ft thick. The Carlsbad Potash District is shown in cross-hatch pattern. Source: Barker and Austin (1995).

Within part of the Permian, the Ochoan, during which the sea was blocked intermittently, the following four groups of rocks present at the WIPP site were formed successively:

1. Castile Formation,
2. Salado Formation,
3. Rustler Formation, and
4. Dewey Lake Red Beds.

Castile Formation

The oldest and deepest of these rock units is the Castile Formation. Castile rocks are composed mainly of anhydrite with alternating thin limestone layers and a few thick layers of halite. Anhydrite and/or gypsum begins to precipitate from brines when 75 percent of the original amount of seawater has evaporated, continuing to be deposited until 90 percent has evaporated, at which point halite begins to precipitate (Kay and Colbert, 1965, p. 202). The alternating character of the Castile suggests that limited amounts of fresh seawater were able to enter the basin intermittently.

Salado Formation

The Salado ("salty" in Spanish) Formation, in which the WIPP facility is located at 658 m below the surface, is composed predominantly of layers (beds) of rock salt, 200-400 m thick. It contains very thin layers (typically 0.1-1 m thick) of other materials (such as clay, anhydrite, and potash minerals) intercalated throughout the formation. Because these evaporite beds have been undisturbed by tectonic forces (large-scale movements of the earth's crust) for approximately 240 million years since their original accumulation, they are essentially horizontal and therefore can be traced continuously for great distances. The thin, distinctly identifiable nonhalite interbeds, called "marker beds" (MBs), have been numbered for identification from the lowest, at the top of the formation, to the highest, near the bottom. Marker beds have been used in the construction of the WIPP facility to keep the floor at a specific stratigraphic level, about 660 m below the surface. The floor of the facility is just above anhydrite MB 139 (Figure A.2).

About midway between the bottom and top of the Salado Formation is a notably thick series of beds, known as the McNutt member of the Salado, which contains economically valuable amounts of potash. The McNutt member, stratigraphically above the salt in which WIPP is located (Figure A.2), is about 120 m thick (Griswold, 1995) but is limited in areal extent within the evaporite basin (Figure A.1). The potash minerals, however, are of sufficient quantity for mining and extraction to have been a source of major economic activity in the area for many years.

Rustler Formation

Overlying the Salado Formation is the Rustler Formation, a thinner-bedded series of strata consisting of five distinct subunits (such sequences are called "members" of a formation). At the bottom of the formation are nonevaporite beds derived from clay muds and sands alternating with layers of halite, anhydrite, limestone and dolomite, and gypsum. The total thickness of the Rustler is about 100 m. The presence of nonevaporite materials such as the clay, silt, and sand (called "clastic deposits" or "clastics") indicates that periodic fluctuations in basin water levels and intermittent inflow of river and marine waters carried clastics into the basin material that sometimes formed separate layers without evaporites, such as sandstones and shales, or settled with the evaporites. The Culebra Dolomite Member of the Rustler Formation, which lies directly above the lowest Rustler clastic strata, is only about 7-8 m thick. The chemical, mineral, and hydrogeologic characteristics of the Culebra are important to a ground-water flow model for the WIPP area.

Dewey Lake Red Beds

The youngest (uppermost) Permian strata in the WIPP vicinity are thin, reddish beds of clay, silt, and sandy sediments produced in a restricted marine environment. The red color derives from the oxidized iron dispersed throughout the clay and sand, suggesting semiarid conditions during deposition of the sediments.

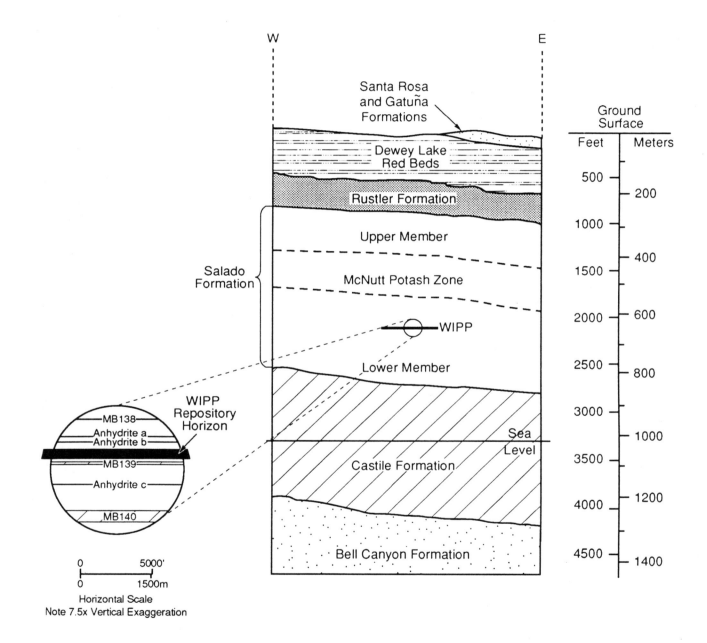

FIGURE A.2 WIPP stratigraphy and depths of four key formations (Castile Formation, Salado Formation, Rustler Formation, and Dewey Lake Red Beds), including the position of the WIPP repository within the Salado. Inset shows finer-scale stratigraphy around the repository horizon, with marker beds and other thin beds. Source: Jensen et al. (1993).

Post-Permian Rocks

An assortment of rock types and ages younger than Permian is present in different locales around the region. Some discontinuous surface deposits were carried by high-energy streams that formed from the melting glaciers to the north during the Ice Age (the Pleistocene, from about 1.6 Ma to 10,000 years ago). One of the Pleistocene stream channels is still present west of the WIPP area as a broad, shallow valley, called Nash Draw, that is wider than 6 km in places (Figure A.3). The valley formed when the high-energy streams eroded down through the rock layers and exposed parts of the Rustler Formation, which are now visible in the walls of the valley. Nash Draw plays a role in determining the regional and local hydrology in the vicinity of WIPP.

HYDROLOGIC SETTING OF WIPP

Because flowing ground water is one of the most likely means by which contaminants could reach the accessible human environment, it is important to understand the hydrogeologic character of the region.

Most of the hydrologic studies of the WIPP site and surrounding region have focused on the Rustler Formation in general, and on the Culebra Dolomite Member in particular. Because the latter is a water-bearing unit that is exposed in Nash Draw, it is a major concern in assessing the capabilities of WIPP to isolate radioactive wastes. However, the other major formations also play a role in WIPP hydrology for varying reasons.

The top of the Castile Formation lies about 200 m below the level of the Salado Formation at which the WIPP facility is located (Figure A.2). Under natural conditions, there is no flow of ground water between the Castile and Salado Formations. However, during exploratory drilling to locate a site for WIPP, project scientists found large, highly permeable zones (pockets) at depths within the Castile, filled with brine under very high pressure (from the weight of hundreds of meters of overlying rocks). This is considered significant in evaluating the WIPP facility because brine in pressurized pockets, once penetrated by drilling, flows out to the surface rapidly and could continue to flow for hours or days, depending on the depth of the pocket, the volume, and pressure of the brine.

Hydrogeology of the Salado Formation

The mechanical behavior of salt, especially massively bedded salt, provides the basis for the inference that water cannot flow continuously through salt along interconnected pathways as it does in many other types of geologic materials. As a weak solid, salt flows under low pressure after a period of time, much as glacial ice flows.

Both are solids, but under pressure such as its own weight or the weight of material above, the crystal structure of the individual minerals that make up the rock salt deforms, causing movement within the minerals (intracrystalline gliding) and movement between crystals (intercrystalline slip) over time. This type of movement, termed "creep," is a very slow process in human terms, in which the actual movement of the salt cannot be seen, although the results of creep are readily observable in room closure and the deformation of boreholes.

The continuous creep of rock salt causes opening and closing of pore spaces that trap droplets of water and prevent the continuous flow of ground water through the interconnected pore spaces in the rock from one location to another. Creep also closes fractures and "heals" them as salt crystals flow into them and interconnect with other salt crystals. For this reason, the permeability of salt (the degree of interconnected pore spaces that allows flow of ground water through the rock matrix or fractures) is generally regarded as extremely low to zero. Pressure tests done on the Salado Formation at various times, both from wells drilled at the surface and from underground tests after the facility was constructed, have indicated that the salt permeabilities are very low, close to the sensitivity limits of the instruments (Mercer, 1987; Beauheim et al., 1991, McTigue, 1993).

During the construction phase of the WIPP facility, brine was observed flowing or seeping into the space created whenever a new opening was made in the salt, whether in a narrow hole or in a large room. These observations led to consideration of the possibility that the disposal rooms could be flooded over the time required for isolation of the long-lived radionuclides

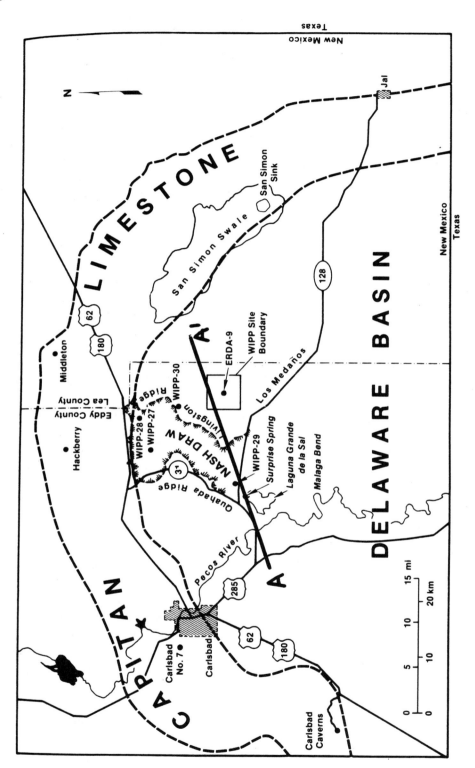

FIGURE A.3 Position of Nash Draw with respect to WIPP. Other features of interest are the location of the Pecos River and the outline of the buried Capitan reef limestone, which encloses the Delaware Basin on the north. Source: Siegal et al. (1991).

and that the consequent chemical reaction of the brine with metal containers would produce large volumes of hydrogen gas (Brush, 1994). These considerations assumed that brine was flowing in from the far field and had a continuous source. However, tests to monitor and measure the inflow of brine over a period of years have shown that the rate of inflow is greatest immediately following the disturbance of the salt to produce a new opening, then declines rapidly, and finally tapers to almost immeasurable amounts (Deal et al., 1991a, b).

These observations have led WIPP project researchers to consider that brine inflow into a newly created opening results when the salt is disturbed by drilling or mining and fracturing occurs in a zone around the new opening (disturbed rock zone, or DRZ), thus releasing brine from entrapped pore spaces (Stormont et al., 1991). Further details and discussion of this issue can be found in Chapter 3 and Appendixes C and D.

The only likely source of measurable Darcy flow in the Salado is the nonhalite interbeds, especially the anhydrite interbeds, which show varying degrees of fracturing. Fracture permeability can provide interconnected pathways for ground water.

Hydrogeology of the Culebra Dolomite

The geological units comprising the Rustler Formation are illustrated in the stratigraphic column shown in Figure A.4. Note that the lithologies represented in this figure are generalized representations for the region and do not necessarily depict the exact distribution of rock types encountered in the Rustler at the WIPP site itself. As illustrated in Figure A.4, the Rustler outcrops are in the Nash Draw area several miles west of the WIPP site, but the formation lies at a depth of about 200 m below the land surface at WIPP.

Ground-water flow in the Rustler Formation is restricted mostly to the Rustler–Salado contact zone, the Culebra Dolomite Member, and the Magenta Dolomite Member (Brinster, 1991). Of these, the Culebra is the most transmissive. Although it is relatively thin (the thickness of the Culebra is generally between 7 and 8 m), it is also extensive in area.

The Rustler Formation, especially the Culebra Dolomite Member, has been the principal focus of hydrogeologic characterization of the WIPP vicinity (Lappin, 1988). The ground-water system in the Rustler is generally characterized by relatively low permeability (or transmissivity) and relatively poor water quality (i.e., too salty to be potable). The Rustler contains a predominance of secondary permeability features, such as fractures, bedding planes, and dissolution features; these may dominate flow in parts of the system, complicate analysis, and render quantitative descriptions difficult. At the very least, these features imply that the ground-water flow system is three-dimensional in a heterogeneous framework that is a challenge to model and understand properly. Both observation and water-supply wells are relatively scarce, contributing to the difficulty and expense of characterization. Although much excellent work has been done during the past 20 years in an attempt to characterize the relatively complex hydrogeology of the area near the WIPP site, some large uncertainties remain in the understanding of the subsurface system above the Salado. The recent state of knowledge has been summarized by Brinster (1991).

A number of water-level measurements and observations of drawdown and/or recovery during hydraulic tests have been used to measure the hydraulic heads and transmissivities in the Culebra. The changes in head over distance determine the hydraulic gradient. Transmissivity, which can vary significantly in value from point to point, is a measure of the ability of the rock to let water flow through it under a given hydraulic gradient.

However, some inconsistencies remain in the interpretation of the long-term flow history of the Culebra based on hydrologic evidence, compared to that inferred on the basis of geochemical and isotopic indicators. These inconsistencies, which reflect some degree of lack of understanding of present or past flow regimes within the Culebra, mean that there is uncertainty in the conceptual model that underlies the performance assessment (PA) model.

Ground-Water Flow Directions

In general, the direction of ground-water flow in the Culebra Dolomite in the WIPP area is from north to

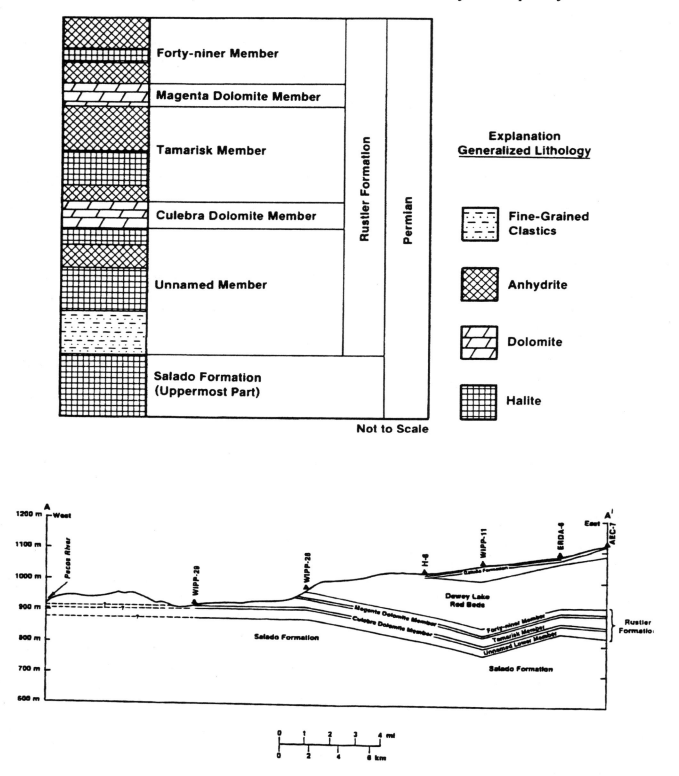

FIGURE A.4 General stratigraphic column showing the five members of the Rustler Formation (top), and cross section from A to A' of Figure A.3 showing Rustler Formation stratigraphy (bottom). Source: Brinster (1991).

south. Because the salinity of the ground water is high and variable, the fluid density also varies significantly. Davies (1989) analyzed variable density flow in the Culebra and concluded that, at least in areas south of WIPP, flow directions can be calculated accurately only if variations in fluid density also are evaluated. Potentiometric data in the Magenta Dolomite indicate that the flow direction is predominantly from east to west. Differences in flow directions within the Rustler Formation reflect the complexity of the hydrogeologic system and may be influenced by vertical components of flow.

A fair degree of uncertainty remains about the amounts and locations of recharge to and discharge from the Culebra. In general, Brinster (1991) notes that recharge occurs north of the WIPP site, although some recharge may also occur east of WIPP because of vertical leakage from the Magenta. Natural discharge is south of WIPP, with some discharge to the Pecos River likely occurring at Malaga Bend.

Transmissivity Estimates

Transmissivity estimates for the Culebra Dolomite in the WIPP area and in Nash Draw range over more than six orders of magnitude. Transmissivity is lowest east of WIPP ($\sim10^{-10}$ m^2/s) and tends to increase to the west ($\sim10^{-3}$ m^2/s). However, the transmissivity is highly variable and spatially correlated (e.g., LaVenue et al., 1995). In some areas, the transmissivity is controlled largely by fracture permeability, whereas in other areas it is governed by primary intergranular properties.

Travel Time

The time it takes for a parcel of water in the Culebra Dolomite to travel from the center of the WIPP site to its southern boundary is an important characteristic of the system controlling the transport of any contaminants released into the Culebra. This travel time depends on the transmissivity, hydraulic gradient, and effective porosity. Analyses assuming that the effective porosity governing transport is a value characteristic of the rock matrix (about 0.16) indicate that the mean travel time would be of the order of 14,000 years (LaVenue et al., 1990). However, if one assumes that the effective porosity is characteristic of

fracture porosity ($\sim$0.0015), then the travel time is reduced to only about 200 years (Davies et al., 1991).

Hydrogeology of the Dewey Lake Red Beds

The Dewey Lake Red Beds are believed to be less permeable than the Culebra Dolomite. However, very few hydraulic tests of this unit have been completed, and few observation wells are available to characterize the flow system. Brinster (1991) reports that although a continuous saturated zone has not been found, some localized zones of relatively high permeability have been identified. Potable water has been reported in some parts of Dewey Lake, perhaps because it is sufficiently shallow to receive direct recharge from precipitation events.

NATURAL RESOURCES

The geologic evolution of this region has resulted in the presence of economically valuable natural resources. The late Permian evaporating basin accumulated a large deposit of potash. The region in earlier Permian time can be characterized as a marine shoreline and some deeper offshore basins accumulating sands and limestone, clastics, and organic debris that resulted in accumulations of economically recoverable oil and gas in rock strata below the evaporites. Thus, the two major activities to extract mineral wealth from the WIPP region are the mining of potash and the drilling and extracting of oil and gas.

Potash Mining

Potash, the commercial name for potassium-bearing minerals, is mined for its potassium, one of the three main plant nutrients in fertilizer. The potassium minerals are concentrated in a middle zone of the Salado Formation, the McNutt Member, which is about 140 m thick (Figure A.2). The Carlsbad Potash District contains the largest potash reserves in the United States (Barker and Austin, 1995; see also Figure A.1). The district is bounded on the west by the dissolution of the shallow Salado strata caused by circulating ground water in the Pecos River drainage area to the west and south of Nash Draw.

Potash mining has been going on in the WIPP vicinity for more than 60 years. Potash was discovered in southeastern New Mexico in 1925. The first mine shaft to the potash 324 m (1,062 ft) below the surface was completed in 1930, and the first commercial shipment took place in 1931. Over the next 30 years, development and production advanced steadily as a large number of companies grew and merged. By the 1940s, New Mexico was the largest domestic potash producer, supplying 85 percent of the country's consumption.

This growth came to a halt and started to decline when potash was discovered, produced, and exported to the United States from Canada in the early 1960s at a lower cost than was economically feasible for U.S. producers. By 1970, Canadian imports exceeded domestic production. The decline continued despite a favorable ruling of dumping against Canadian producers by the International Trade Commission in 1987. Shortly thereafter, the former Soviet Union countries began exporting potash to the United States, further increasing competition and depressing both price and demand. Because of the foregoing and other economic and declining resource factors, the potash industry in the WIPP region is now much depressed, and very few companies are still active (Barker and Austin, 1995).

The mining front is now close to the WIPP site boundary and has reached the edge of the study area on the southwest side of the WIPP site (Figure A.5). In the future, the southwest or north side of the WIPP boundary may be the next target area for mining (Griswold, 1995).

In assessing potential future impacts of potash mining on the WIPP facility, the adoption of solution mining of potash minerals, a technique now being used in Canada, has been mentioned as a possible development. However, of the two dominant potash minerals being mined in the Carlsbad Potash District, sylvite and langbeinite, only sylvite is sufficiently soluble to make this mining process economically and practically feasible. According to specialists familiar with potash mining in this area, the beds containing sylvite are considered too thin for solution mining to be practical, but mine owners have not discounted this process as a future development in the area (Griswold, 1995).

If the potash resources overlying the repository were mined in the future, the resulting subsidence could significantly increase the transmissivity of the Culebra Dolomite, located approximately 200 m above the potash (Figure A.2). This could result in faster release of radionuclides to the environment following a human intrusion event.

Oil and Gas Resources

After the strata below the evaporites of the Delaware Basin had been bypassed for many years as unlikely to be producers of economically recoverable oil resources in the WIPP area, oil was discovered in the late 1980s and early 1990s. Indications in earlier exploration had suggested that mostly water was present in these strata. New techniques in analyzing the geophysical logs led to identification of several commercial oil pools in the Delaware Mountain Group of rocks at and below about 2,500 to 2,700 m in depth.

Major oil drilling activity began in the Delaware Basin following the discovery. This is now one of the most active areas of oil exploration and extraction in the United States (Broadhead et al., 1995, p. XI-10; Figure A.6). In the vicinity of WIPP, four major pools have been discovered, at least one within 1 km of the site boundary.

New technologies in oil drilling have improved recovery, safety, and cost efficiency. Improvements in well productivity result from advances in directional drilling, completion technologies, and stimulation techniques (Hareland, 1995). These advances and projected future oil and gas production were used to calculate the estimated value of oil and gas underneath the WIPP land withdrawal area. The estimated total primary and secondary recovery of probable resources in oil reservoirs under WIPP is about 54 million barrels of oil and gas condensate (Broadhead et al., 1995).

The secondary oil recovery technique known as waterflooding has been used in some areas around WIPP. A concern has been raised that such a process could introduce large amounts of water into the Salado Formation that could find its way to the WIPP facility by flow through marker beds, a process that has been observed in some oil fields in the region. In 1991, a flow of up to 1,200 barrels per hour of brine inflow was encountered in a hole being drilled for petroleum production. At the time of inflow, the hole was approximately 100 m above the base of the Salado. The inflow was attributed to a waterflooding operation in the Rhodes Yates Field in which a large number of old wells were simultaneously injected. These injections occurred in the Yates Formation some 100 m below the base of the Salado, and 2 miles (3 km) from the well where the inflow occurred. This has raised concerns that similar inflows may occur due to waterflooding operations near WIPP and could compromise the integrity of the repository. This is further discussed in Chapter 3 (Box 3.2). See Figure A.7. A full discussion of waterflooding is presented in Broadhead et al. (1995).

A key issue is whether a future society would be aware of WIPP and its hazardous contents. The presumption that the existence or significance of WIPP would be lost is subjective and unquantifiable and therefore beyond the realm of rigorous scientific prediction. A second issue is whether such a society would elect to drill through it if knowledge of WIPP's existence were retained.

If future societies retain a knowledge of the existence of WIPP, yet still want to explore for or recover petroleum resources below WIPP, then directional drilling is feasible and likely. With present technology, a drilling operation could avoid damage to the repository if it were operated from the surface beyond the WIPP site boundary, and directed to avoid the waste horizon.

Manmade site markers, to be emplaced after closure, are required under the EPA standards. The reason for this requirement is that such markers, if they persist over time and are understood, could reduce the likelihood of inadvertent intrusion. The intent of erecting markers is to lead drillers to reconsider vertical boreholes that would penetrate the repository, and to choose instead alternative methods such as directional drilling techniques.

The assumption that the WIPP area would be drilled by today's methods for the next 10,000 years is also questionable. The recent advances in directional drilling techniques show that technology evolves in time.

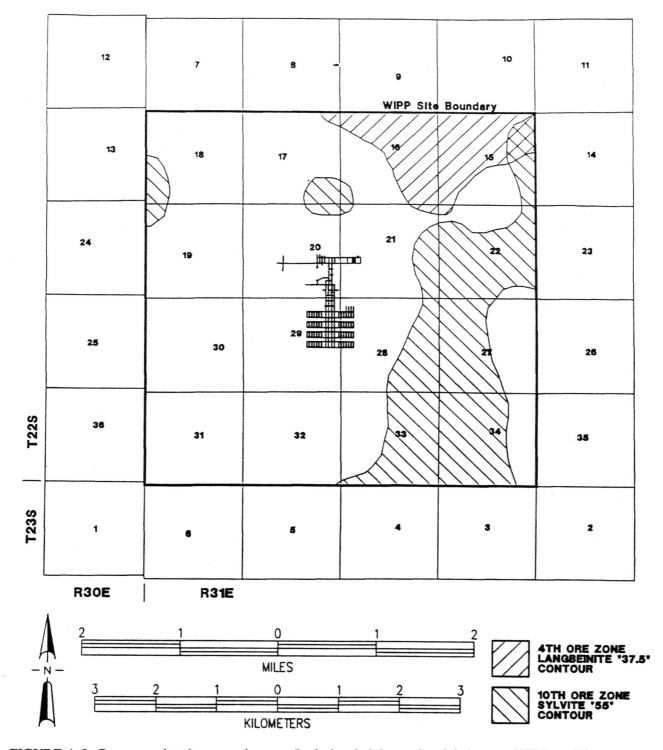

FIGURE A.5 Contours showing potash ores (both langbeinite and sylvite) near WIPP. (The contours are interpolated from figures 4 and 30 of Griswold [1995]). Also shown is the location of the WIPP facility of Figure ES.1.

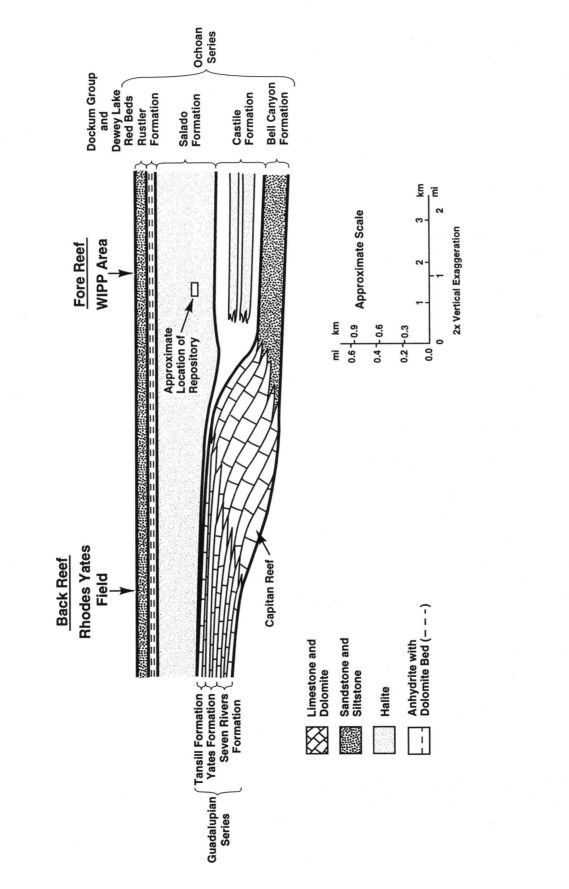

FIGURE A.6 Cross section depicting the relative locations of the Rhodes Yates Field and the WIPP repository.
Source: Sandia National Laboratories, unpublished. Data from the State of New Mexico Bureau of Mines.

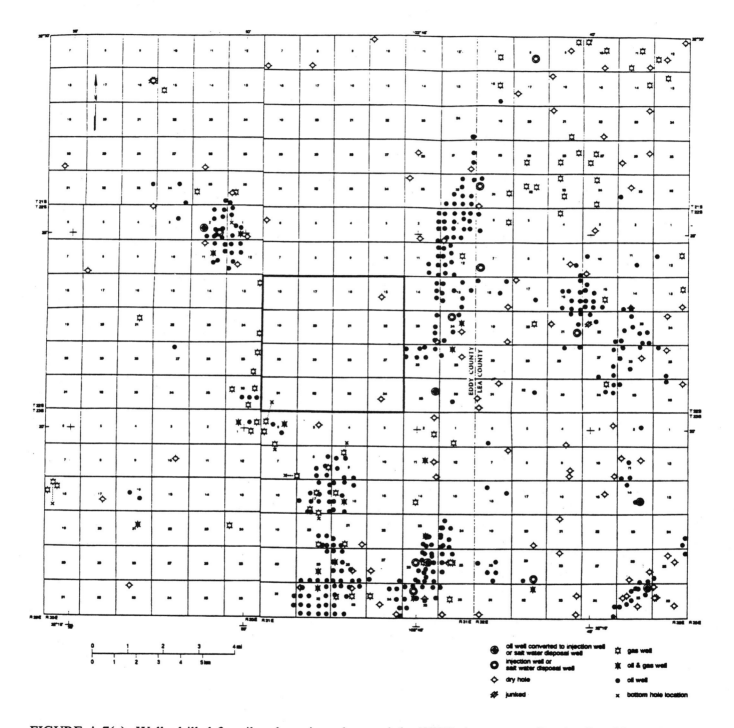

FIGURE A.7(a) Wells drilled for oil and gas in and around the WIPP site, surrounding 1-mile-wide study area, and nine-township project study area. Note that some wells are right at the WIPP boundary. No producing wells are within the WIPP exclusion zone. Source: Adapted from Griswold (1995).

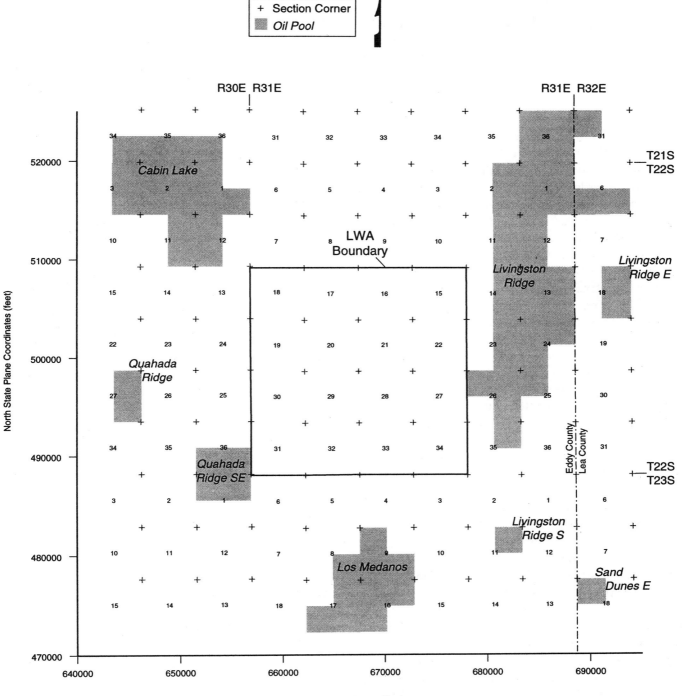

FIGURE A.7(b) Oil pools near the WIPP site, outside of the Land Withdrawal Act boundary, shown on state plane coordinates. Note that two pools abut the LWA boundary. Source: Sandia National Laboratories, unpublished. Data from the Petroleum Institute.

Appendix B

The Complementary Cumulative Distribution Function: The Risk Curve

Risk assessment, including performance assessment, has created the ubiquitous complementary cumulative distribution function (CCDF). Although some advocate a less imposing label such as "the risk curve," CCDF seems to have found its place in the risk literature as the preferred name. As shown below, the words do have mathematical meaning.

What is a CCDF? One answer is that the CCDF is an aggregated response to the triplet definition of risk noted in Chapter 2 of this report. In particular, for the Waste Isolation Pilot Plant (WIPP) performance assessment, the CCDF consists of plots on log-log graph paper of exceedance probability versus consequence. There may be several CCDFs to cover several different consequences of interest. There may also be several CCDFs for a particular consequence to indicate the range of uncertainty involved.

Most people are familiar with the concept of a bell-shaped curve as a way to convey confidence, or probability, in the value of a parameter, such as the number of curies released from an inventory of radionuclides. Such a curve tells how much of the probability is associated with intervals of curies released. This curve, called the probability density function, is the probability per unit interval of curies released. Of course such curves can be discrete, as in a histogram, or smooth, as in a continuous function (see Figure B.1[a]).

A more interesting question than the probability per release interval is referred to in risk assessment as "the exceedance question." This is the probability that the release will exceed a certain value (in the above example, exceed a certain number of curies of a particular radionuclide). This question can be answered by a summing, or integration operation, on the probability density function (Figure B.1[b]). The

result of such a summation is called the cumulative distribution function. The complement—that is, one minus the parameter (here, the cumulative probability)—and the log-log scale are the additional steps taken to achieve the desired form (Figures B.1[c] and [d]). These steps result in a compact form for representing parameters that cover an extremely wide range of values.

Suppose, in the spirit of the triplet definition of risk, that a performance assessment has been conducted and a set of scenarios has been developed, each with its own probability density function of the number of curies of a particular radionuclide released. To cast the results in complementary cumulative form, the scenarios are structured in order of increasing release fractions and the probabilities are cumulated from the bottom to the top as a function of the different release fractions. Plotting the results on log-log graph paper generates a curve of the form shown in Figure B.1(d).

Figure B.1(d) does not represent reality, because it suggests that the outcome of a consequence is known with complete certainty. In practice, there are no absolutes; rather, there is significant uncertainty, starting with the uncertainties of the many individual inputs that are used to calculate a CCDF for a typical risk assessment. Thus, it is impossible to specify a single CCDF as the unambiguous outcome of a risk assessment. Instead, there are many possible CCDF curves (Figure B.2), each with its own likelihood of being correct. The spread of these CCDF curves provides a measure of the confidence with which the CCDF, for the consequence under consideration, can be estimated.

The uncertainty that gives rise to the CCDF curves in Figure B.2 is typically characterized by developing a probability distribution for each imprecisely known

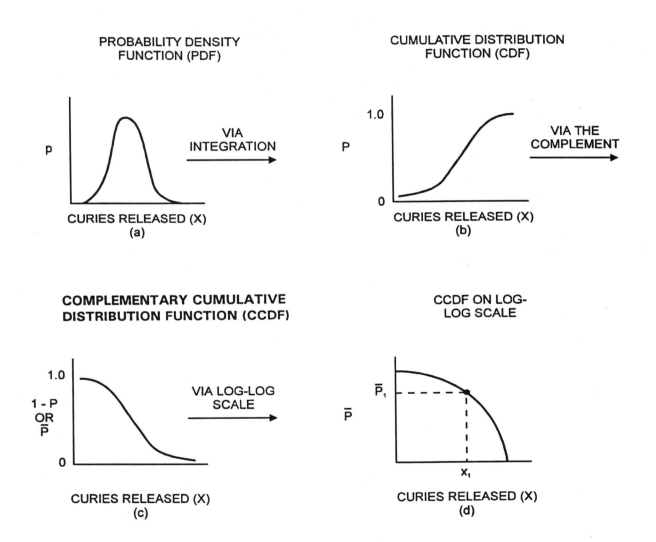

FIGURE B.1 This figure provides an exact interpretation of the parameters $\overline{P}$ and X. In particular, the CCDF representation says that the probability $\overline{P}$ of releasing x_1 curies, or more, does not exceed $\overline{P}_1$.

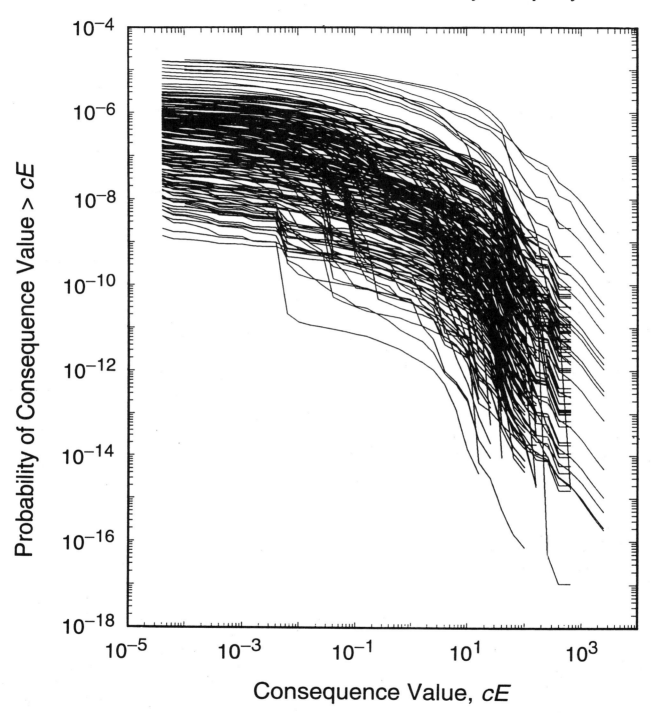

FIGURE B.2 Example distribution of CCDF curves obtained by sampling imprecisely known analysis inputs. Source: Helton et al. (1991).

input used in the analysis. These distributions mathematically describe a degree of belief, based on all the available evidence (e.g., data, background knowledge, analyses, experiments, expert judgment), of the range and weight, in terms of likelihood, of the input values used in the analysis. An example of an input might be a distribution coefficient for a radionuclide transport calculation. Such distribution coefficients cannot be assigned a fixed number because of the uncertainty of what that number should be. Thus, the input has to be in the form of a probability distribution that expresses the analyst's state of knowledge about what the number should be. It tells a much fuller story than would be provided by a single number. The CCDF curves in Figure B.2 arise from the distributions assigned to the individual inputs.

The usual computational procedure is to generate a sample (e.g., random or Latin hypercube[1]) from the uncertain inputs according to their assigned distributions and then to construct one CCDF of the form appearing in Figure B.2 for each element of this sample. With this procedure, each CCDF is constructed from a set of input values (sometimes referred to as a "realization") that is consistent with all available information. Further, the assignment of distributions to individual inputs, and the propagation of these distributions through to the distribution of CCDF curves in Figure B.2, provide a representation of the uncertainty in the final outcome of the risk assessment, where this outcome is a CCDF. Put another way, the distribution of CCDF curves in Figure B.2 provides a measure of the confidence with which the outcome of the risk assessment can be estimated. A tight grouping of CCDF curves in Figure B.2 indicates a high confidence in the estimated location of the CCDF of interest; conversely, a wide spread in the CCDF curves indicates a low confidence in the estimated location of this CCDF.

Although Figure B.2 provides a complete summary of the distribution of CCDF curves obtained by propagating the uncertainty in the individual analysis inputs, it is rather crowded and difficult to read. A less crowded summary can be obtained by plotting the mean value and selected percentiles for each consequence value on the abscissa. For example, the mean plus the 5th, 50th (i.e., median), and 95th percentile values might be used (Figure B.3). The percentile curves then provide a measure of confidence as to the location of the exceedance probabilities for individual consequence values (e.g., there is a degree of belief probability of .9 that the exceedance probability for a particular consequence value is located between the 5th and 95th percentile curves). Further, the mean CCDF, obtained by vertically averaging the individual CCDF curves in Figure B.2, provides a measure of central tendency; the 50th percentile curve provides a related, but different, measure of central tendency. However, it is the distribution of CCDF curves that provides the overall measure of confidence that can be placed in the results of the risk assessment.

The distribution of CCDF curves in Figure B.2, and hence the mean and percentile curves in Figure B.3, are typically obtained by sampling-based techniques and therefore are approximate. However, if a robust sampling procedure is used, these results should show little variation from one sample to the next (Figure B.4). In concept, the mean and percentile curves in Figure B.3 can be specified uniquely by approximately defined integrals; in practice, these integrals have to be approximated by sampling-based techniques.

---

[1]Latin hypercube sampling is a sampling scheme in which the probability distributions of the input variables are discretized into intervals of equal probability. The advantage of Latin hypercube sampling is claimed to be greater accuracy due to the more detailed sampling of regions of higher probability.

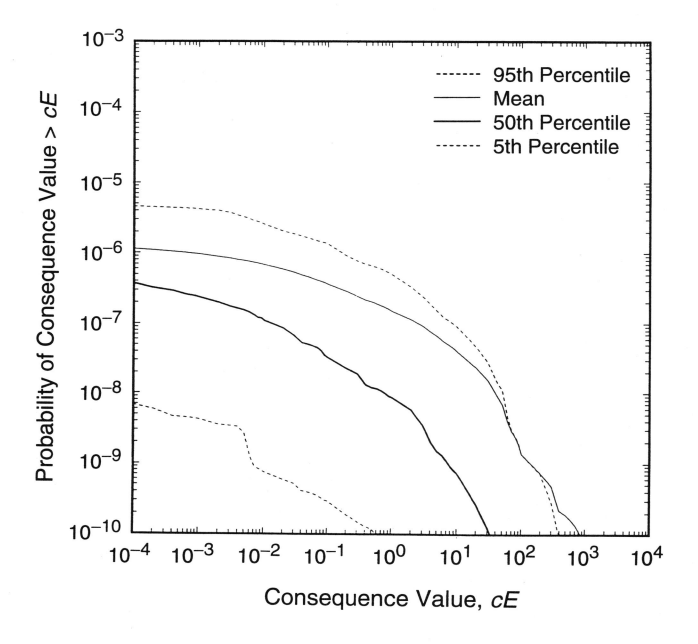

FIGURE B.3 Example summary curves derived from an estimated distribution of CCDF curves. The curves in this figure were obtained by calculating the mean and the indicated percentiles for each consequence value on the abscissa in Figure B.2. Source: Helton et al. (1991).

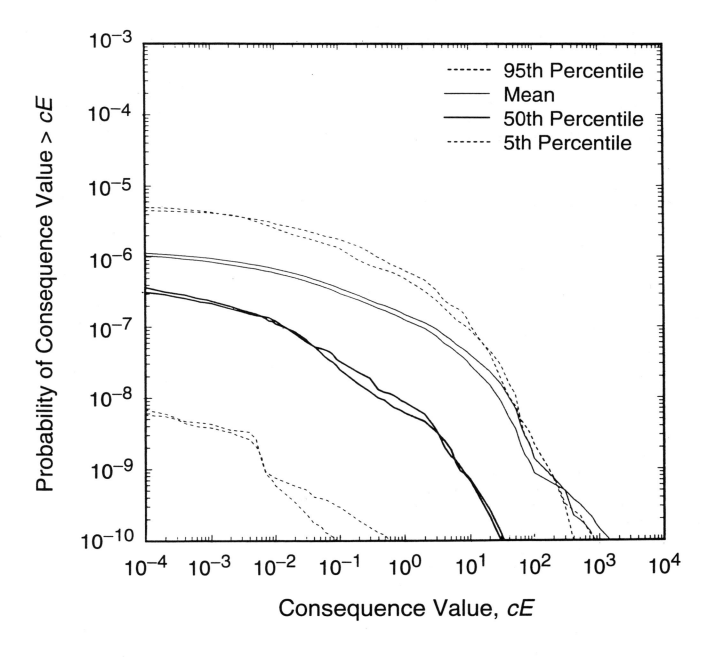

FIGURE B.4 Example of mean and percentile curves obtained with two independently generated samples for the results shown in Figure B.3. Source: Helton et al. (1991).

Appendix C

Brine Inflow to Excavations in the Salado

PERMEABILITY OF WIPP SALT ANHYDRITE AND INTERBEDS

Observation of brine seeps into excavations (both tunnels and boreholes) at the WIPP has led to the assumption that the Salado salt has a finite, but extremely low (~ 10^{-21} – 10^{-23} m^2) permeability in the classical sense of Darcy flow. There are several inconsistencies in this model. An alternative model in which undisturbed salt is assumed to contain brine only in unconnected porosity appears in better agreement with underground observations at WIPP. In this model, "damage" to the salt (i.e., development of the disturbed rock zone, DRZ) resulting from introduction of the excavation, leads to the development of a local zone of interconnected porosity characterized by a finite permeability. It is the release of the brine in the previously unconnected pores by the stress-induced damage that produces the brine seeps—at least in the initial stages (Borns, 1995).

McTigue (1995a,b) has examined this alternative model in detail. he notes:

> Data for flow to boreholes in WIPP Room D are represented well qualitatively by the classical model. However, the capacitance indicated by fitting model calculations to the field data is orders of magnitude larger than that expected on the basis of the compressibilities of salt and brine. That is, the decay of the brine flux into the boreholes takes place over a time scale that is much too long to be explained in terms of the processes assumed in the development of the classical model. The classical model invokes an unbounded domain of interconnected porosity. This concept is contradicted by both mechanical arguments and

> geochemical observations. (McTigue 1995b).

The mechanical arguments from the same reference point out that salt creeps at vanishingly small deviatoric stress (i.e., stress difference). Thus, over long time frames, interconnected porosity can not be sustained. Brine will be confined to local domains of isolated porosity containing fluid at a pressure equal to (minus) the mean [i.e., lithostatic] stress in rock.

Geochemical observations, however, indicate that

> ...brine chemistry is highly variable, even among samples separated by distances of the order of tens of centimeters. If these brines were derived from an interconnected pore network, one would expect that molecular diffusion would have eliminated any significant contrasts in brine composition over the very long existence of the formation....Over a period of 230 Ma the diffusion length, L_a, is...of the order of hundreds of meters. The observation that compositional differences persist in the brines over short length scales and very long time, then, suggests strongly that the brine in undisturbed salt is local, isolated domains. (McTigue, 1995b)

McTigue (1991) later analyzes brine seepage rates into Room Q and again concludes that this experiment confirms the view that undisturbed salt is impermeable.

Borns (1995) notes that measured changes in the electrical resistivity in the DRZ around Room Q indicate an initial desaturation (as the DRZ develops) followed by a resaturation over a period of approximately three years. This could suggest that Darcy flow may occur in the salt, but it could also be the result of the progressive development of the DRZ, which eventually connects the anhydrite layers hydrologically to the room excavation.

The McTigue and Borns observations can be reconciled by a model in which classical Darcy flow occurs in the anhydrite marker beds (e.g., MB 139), but the salt between these layers is impermeable.

The following calculations indicate the order of magnitude of brine inflow predicted on the basis of assuming

1. an impermeable salt containing anhydrite interbeds that exhibit Darcy flow;
2. salt with a small, but finite, permeability, allowing classical Darcy flow radially into a circular excavation. (In this case, the permeable interbeds are assumed to be an integral part of the homogeneously permeable salt.)

ONE-DIMENSIONAL FLOW IN ANHYDRITE INTERBEDS IN IMPERMEABLE SALT

The instantaneous rate of fluid inflow q, at some time t, per meter of tunnel length[*] from a horizontal permeable layer (Figure C-1), thickness d and of infinite extent (one-dimensional flow) can be calculated (Churchill, 1972) from the formula shown in (C-1):

$$q = kd\frac{\phi_0}{\sqrt{\pi ct}} \qquad \text{(C-1)}$$

where $k = \kappa pg/\mu$ is the brine conductivity, d is the thickness of the layer, ϕ_0 is the initial head, and $c = k/S_s$ is the diffusivity. (Other symbols are defined in Tables C-1 and C-2.) In the case of a constant head at a finite distance L from the tunnel along the layer, the instantaneous rate of inflow can be written (Churchill and Brown, 1987) in dimensionless form:

---

[*] Only one side of the tunnel is considered–the total inflow into the tunnel would be double the values shown in equations (C-1), C-2), and (C-3). The remainder of the rock mass is assumed to be impermeable.

$$\bar{q} = 2\sum_{m=0}^{\infty} \exp\left(-m^2\pi^2\tau\right) - 1 \qquad \text{(C-2)}$$

where $\bar{q} = Lq/dk\phi_0$ and $\tau = tc/L^2$. The dimensionless rates of inflow, from an infinite layer and from a layer with a fixed head at the finite distance L, as a function of dimensionless time, are shown in Figure C-2.

The total inflow Q per meter of the tunnel—one side of the tunnel only—from an infinite layer can be obtained by integrating (C-1):

$$Q = \int_0^t q(t)\,dt = 2dk\phi_0\sqrt{\frac{t}{\pi c}} \qquad \text{(C-3)}$$

The curves shown in Figure C-3 are calculated by using the parameters for an anhydrite layer in salt, as shown in Table C-1. Total inflows (C-1) from an infinite layer for the base case mentioned above and two cases used by McTigue (1991) (parameters shown in Table C-2) are compared in Figure C-4.

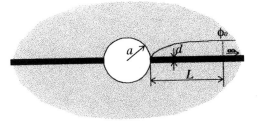

Figure C-1: Model of one-dimensional flow.

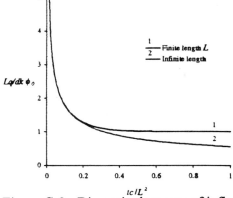

Figure C-2. Dimensionless rate of inflow from one side of a permeable layer.

permeability	κ	10^{-18}	(m^2)
brine viscosity	μ	$1.8 \cdot 10^{-3}$	$(Pa \cdot s)$
specific weight	ρ	1230	(kg/m^3)
initial head	ϕ_0	650	(m)
fixed head distance	L	1000	(m)
layer thickness	d	1	(m)
specific storage	S_s	$1.5 \cdot 10^{-7}$	$(1/m)$
conductivity	k	$6.69 \cdot 10^{-12}$	(m/s)
diffusivity	c	$4.45 \cdot 10^{-5}$	(m^2/s)

Table C-1. Base case values of parameters used to calculate brine inflow into a tunnel from a horizontal anhydrite layer.

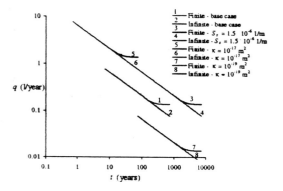

Figure C-3. Rates of inflow from the anhydrite layer (per meter of length along the tunnel). (The rates shown are for one side of the tunnel only, i.e., total inflow rates are double those shown).

	Base case	McTigue case 1	McTigue case 2	
κ	10^{-18}	10^{-21}	10^{-22}	(m^2)
μ	$1.8 \cdot 10^{-3}$	$1.60 \cdot 10^{-3}$	$1.60 \cdot 10^{-3}$	$(Pa\ s)$
ρ	1230	1230	1230	(kg/m^3)
ϕ_0	650	815	815	(m)
d	1	1	1	(m)
k	$6.69 \cdot 10^{-12}$	$7.67 \cdot 10^{-15}$	$7.67 \cdot 10^{-16}$	(m/s)
c	$4.45 \cdot 10^{-5}$	$6.20 \cdot 10^{-8}$	10^{-10}	(m^2/s)

Table C-2. Values of parameters used in "base case" and by McTigue to calculate brine inflow from an anhydrite layer into a tunnel.

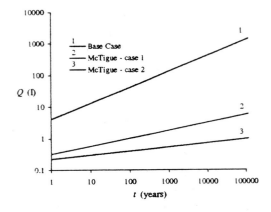

Figure C-4. Total inflow into a tunnel from an anhydrite layer (one side of tunnel only).

RADIAL FLOW INTO EXCAVATIONS IN PERMEABLE SALT

The Laplace transform of the instantaneous rate of inflow $\tilde{q}$ into a circular opening in an infinite, homogeneous, porous medium (Figure C-5) is as follows (Detournay and Cheng, 1988):

$$\tilde{q} = 2\pi a \phi_0 k \frac{K_1(a\sqrt{s/c})}{\sqrt{sc}\, K_0(a\sqrt{s/c})} \qquad (C-4)$$

where K_0 and K_1 are the modified Bessel functions of the second kind, of order zero and one respectively.

Inversion of (C-4) is performed numerically. The total inflow Q (per meter along the tunnel) for $k = 6.25 \cdot 10^{-14}$ m/s (corresponding to $\kappa = 10^{-20}$ m^2), radius of tunnel $a = 1.524$ m, and an initial hydrostatic pressure $p_0 = 6$ MPa, is shown in Figure C-7.

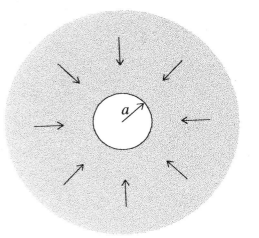

Figure C-5. Model of radial flow into circular tunnel.

BREDEHOEFT CALCULATION (BREDEHOEFT, 1988)

Around 1987, observations of brine inflow into underground excavations at the WIPP led Dr. J.D. Bredehoeft, then a member of the NRC WIPP committee, to suggest that the Salado formation at WIPP could have finite, albeit low, permeability in the classical Darcy sense and that the Salado was possibly saturated with brine. In defining the Salado, he (Bredehoeft, 1988) did not distinguish between the salt and anhydrite interbeds. He calculated, numerically, the daily brine inflow into rectangular excavations for several assumed values of hydraulic conductivity k. [For WIPP brine the intrinsic permeability κ (m^2) = $0.16 \cdot 10^{-6}k$ (m/s). Thus $k = 6.25 \; 10^{-14}$ m/s corresponds to $\kappa = 10^{-20}$ m^2, etc.].

Figure C-6 compares Bredehoeft's results (curves 1-4) with those obtained analytically by using (C-4). It is seen that the two sets of results are consistent. The principal difference is due to the higher range of brine conductivities assumed by Bredehoeft. The values used in the current calculations are based on more recent values used by DOE.

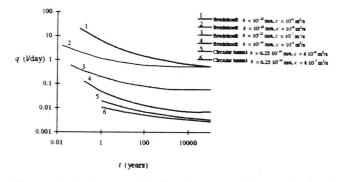

Figure C-6. Rates of inflow into tunnel, as calculated numerically by Bredehoeft (curves 1 to 4) and analytically)(curves 5 and 6).

Using a Darcy flow model to interpret recent in-situ measurements, Beauheim et al., (1993a) conclude that *"the vertically averaged hydraulic conductivities of the tested intervals (primarily anhydrite interbeds) range from about $1 \cdot 10^{-14}$ to $2 \cdot 10^{-12}$ m/s (permeabilities of $2 \cdot 10^{-21}$ to $3 \cdot 10^{-19}$ m^2). Storativities of the tested intervals range from about $1 \cdot 10^{-8}$ to $2 \cdot 10^{-6}$ and values of specific storage range from $9 \cdot 10^{-8}$ to $1 \cdot 10^{-5}$ 1/m. ..."* Anhydrite interbeds appear to be one or more orders of magnitude more permeable than the surrounding halite, primarily because of sub-horizontal bedding-plane fractures present in the anhydrites.

Freeze and Christian-Frear (in preparation) attempting to interpret the brine inflow results for Room Q, indicate that a value of *"far-field permeability of $5 \cdot 10^{-22}$ m^2 with a bulk rock compressibility of $5.4 \cdot 10^{-12}$ 1/Pa"* fits the apparent brine inflow rates from two to five years. Interpretation was complicated by the absence of very early time (0-6 months) brine inflow data—due to malfunctioning of the room seals. Thus, it seems that the hydraulic conductivity of (i) Salado halite (interpreted in terms of Darcy flow model) is of the order of 10^{-13} to 10^{-16} m/s (permeability 10^{-20} to 10^{-23} m^2) and (ii) anhydrite interbeds is 10^{-12} to 10^{-14} m/s (permeability 10^{-19} to 10^{-21} m^2). Since the anhydrite interbeds are relatively thin layers within the halite, an overall value for conductivity (permeability) of the Salado would tend to be at the lower end of the combined anhydrite—halite range (i.e., say 10^{-14} m/s), resulting in the inflows in the range of curves 4, 5, and 6 in Figure C-6.

Figure C-7 shows the total inflow per meter length of excavation for the lower permeability results (curves 5 and 6) shown in Figure C-6. It is seen that after 100 years, a total of about 200 liters (or 0.3 m³) of brine has flowed into the excavation. If a 10-m excavation width (or "width" of waste material) is assumed, this amount of brine would occupy a height of 2 cm over the 10-m width. If the compacted waste is assumed to occupy 70% of the original space (i.e., leaving 30% voids), then the height of brine would increase to (2/0.3) cm, or approximately 6 cm of the height of compacted waste. However, since the DRZ would exist for at least a substantial part of the 100 years, then much of this brine would gravitate into the fractured rock in the DRZ underlying the waste. Brine from MB 139, below the excavation, would also flow into the same fractured rock below the floor. It seems probable, therefore, that much of the brine will not come into contact with the waste but instead will flow down-gradient in the DRZ below the waste. This suggests that with appropriate repository design, rooms could be maintained in a "brine humid" condition rather than a "brine-flooded" condition. According to Brush, gas generation caused by corrosion of the steel waste drums in brine humid conditions is negligibly small.

Should there be a more significant rate of gas generation, pressure in the rooms would tend to rise, thereby reducing brine inflow—and the associated gas generation rate. Although it is possible, under some conditions, for gas to flow out of the room [e.g., along the (finite permeability) marker beds] at the same time as brine flows in, a recent DOE study by Webb and Larson (1996) indicates that such "counter-current" flow is unlikely for the 1° dip of the marker beds at the WIPP until the gas pressure in the excavation is close to lithostatic pressure. Any gas at lower pressure will directly reduce the hydraulic pressure differential and, hence, reduce the brine inflow into the excavation as well. This, in turn, will limit corrosion and the associated gas generation. If the gas pressure reaches (and "attempts" to exceed) lithostatic, the permeability of the anhydrite interbeds would increase substantially since fracturing of the interbeds would occur and

counter-current flow (gas out, brine in) in the 1° updip section of the anhydrite interbed would result in increased brine inflow into the excavation. From curves presented by Webb and Larson (1996) in inflow appears to be or the order of twice the inflow rate observed for the unpressurized (i.e., open) excavation. With the low brine inflows calculated earlier and the associated low gas generation rates, it seems unlikely that gas pressure could approach lithostatic before the waste-filled excavations had closed to their residual (compacted) volumes.

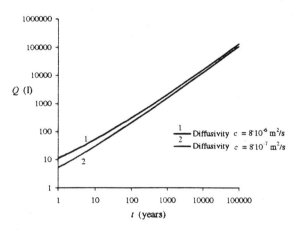

Figure C-7. Total inflow into a tunnel, if radial flow is assumed.

SHAFT SEALS – APPROXIMATE CHECK ON PA FLOW CALCULATIONS

If Darcy flow is assumed,

$$q = -kA\frac{d\phi}{dl} \qquad \text{(C-5)}$$

where q is the volumetric flow rate (m³/s), k is the hydraulic conductivity of the (porous) seal (m/s), dl is the length (m) across which $d\phi$ (m) is measured, and A is the cross-sectional area (m²) normal to the flow direction.

The hydraulic conductivity may be determined from the properties of the porous (seal) material and the saturating fluid. Thus,

$$k = \frac{\kappa \rho g}{\mu} \qquad (C.6)$$

where κ is the intrinsic permeability of the porous (seal) material (m^2); ρ is the fluid density (kg/m^3); g is the gravitational acceleration (m/s^2); and μ is the fluid viscosity[Pa·s].[†]

The term "hydraulic conductance" C (m^2/s) is sometimes used, where

$$C = \frac{kA}{L} \qquad (C.7)$$

and L is the length of the porous material (i.e., the shaft seal in this case).

$$q = -C \cdot \phi \qquad (C.8)$$

where ϕ is the fluid pressure (head) differential (m) over the length L, and

$$v_z = \frac{q}{A} = \frac{C}{A}\phi \qquad (C.9)$$

where v_z is the Darcy velocity[‡] or specific discharge (m/s) of the fluid in the porous medium (seal) with a head differential of ϕ (m).

Using PA assumed values (i.e., a shaft seal length L = 100 m; total area of the four shaft seals A = 100 m^2; seal permeability κ of $1 \cdot 10^{-16}$ m^2; brine viscosity of 0.0018 Pa·s; fluid density of 1230 kg/m^3; and gravitational acceleration of 9.792 m/s^2), yields (U.S. Department of Energy, 1995d, p. 5):

$$
\begin{aligned}
C &= \frac{\kappa A}{L}\frac{\rho g}{\mu} = \frac{10^{-16} \cdot 100}{100}\frac{1.23 \cdot 10^3 \cdot 9.792}{18 \cdot 10^{-4}} \\
&= 6.7 \cdot 10^{-10} \text{ m}^2/\text{s or } 211.3 \text{ m}^2/10^4 \text{ years} \\
v_z &= 2.1 \cdot \phi \text{ m}/10^4 \text{ years} \\
q &= 211.3 \cdot \phi \text{ m}^3/10^4 \text{ years} \\
&\left[10^4 \text{ years} \simeq 3.15 \cdot 10^{11} \text{ s} \right]
\end{aligned}
$$

At the time the seal is first installed, the repository will be essentially dry and fluid flow through the seal initially will be downward—primarily from the Rustler formation overlying the Salado. Flow upward through the seal will begin only when the fluid pressure exerted at the bottom of the shaft seal exceeds the hydrostatic pressure due to the height of water in the shafts.

The predicted change in pressure with time at the shaft bottom will depend on a complex interaction of several repository variables, such as the total amount of void space in the waste-filled rooms and all of the (back-filled or open) excavated space; rate of reduction in void space due to salt creep; rate of inflow of brine into the void space; rate of gas generation and room pressurization; loss in fluid pressure due to fluid flow from the waste-filled rooms to the bottom of the seal, etc. Depending on the particular combination of assumed values of these variables used in a calculation, the brine pressure at the shaft bottom may remain very small for a long time after sealing or may increase after a period of the order of several hundreds of years, when gas generated by brine corrosion of the steel waste containers develops an increasing pressure in the repository.

The relationship between seal permeability and flow through the seal over a given time [Eq. (C.8)] is linear, so the net cumulative flow, over a period of time through a seal of a given permeability can be determined simply from the average driving fluid pressure [i.e., the pressure ϕ above hydrostatic at the bottom of the seal over the same time (This is 650 m for the water-filled shaft.)]. If the pressure ϕ is initially below hydrostatic (i.e., negative), then flow during this period will be negative, that is, downward through the seal.

[†] Pa·s = (N/m^2)·s = (kg$_{mass}$·m/s^2)·s/m^2 = kg$_{mass}$/m·s.

[‡] See discussion of Darcy velocity.

It is seen from (C-8) that the net cumulative flow over 10,000 years through a shaft seal of assumed permeability 10^{-16} m^2 will be 10,000 m^3, if the average (over 10,000 years) pressure head ϕ at the shaft bottom is $\phi = 10,000/211 = 40$ m (above hydrostatic). For a seal of permeability of 10^{-17} m^2, the average pressure head required to produce a cumulative upward flow of 10,000 m^3 over 10,000 years will be $\phi = 470$ m above hydrostatic—(650 m) or a total average brine pressure head of 110 m. Similarly, for a seal permeability of 10^{18} m^2, a total average pressure head of 1120 m would produce a net cumulative brine outflow of 1,000 m^3 in 10,000 years.

Figure 4.1 in Chapter 4 of this report shows the results of a number of calculations—each for a randomly chosen combination of repository parameter values (closure rate, brine inflow rate, etc.)—of the cumulative brine flow through the seals over 10,000 years, as a function of seal permeability. It is seen that, especially at the higher permeabilities, flow *into* the repository (i.e., the negative values) tends to be more likely. For a permeability of 10^{-16} m^2 or lower, the net flow is essentially zero.

Creep of the salt around the shaft will result in further consolidation of the 100-m-high seal of crushed, compacted salt, so that the permeability of the seal will tend eventually towards the permeability of intact salt. As noted above, intact pure salt likely to be impermeable. Where intact salt is assumed to have a finite permeability, this is usually taken as about 10^{-21} ~ 10^{-23} m^2. Calculations indicate (Callahan et al., 1996) that a seal of crushed salt at a depth between 550 m ~ 650 m, initially compacted to a permeability about 10^{-16} m^2 is likely to achieve a permeability of 10^{-20} m^2 or so within 100 years after placement.

Thus, it appears that seals consisting of 100-m-high layers of compacted crushed salt should provide an effective barrier to release of radionuclide contaminated brine to the accessible environment over 10,000 years or longer.

The height to which the brine would rise above the seals depends on the average porosity of the material in the shaft (see discussion of Darcy flow below). Thus, if the shaft above the seal was left completely open and the (total) cross-sectional area (of the shafts) was assumed to remain constant at 100 m^2, a cumulative brine flow of 10,000 m^3 would create a brine column 100 m high above the seal. If the average porosity of the material in the shaft was 1%, the brine would rise, hypothetically, 100 times higher. In the case of WIPP, the brine would rise to the top of the Salado and then would be able to flow laterally into the Culebra.) In both cases, the *total* volume of brine outflow would be essentially unchanged. For a given concentration of raduonuclides in this same volume of brine, the radionuclide release will also be the same, independent of the porosity of the material through which it flows.

Darcy Velocity

The Darcy velocity (or specific discharge) is the volumetric rate of flow per unit area through which flow occurs. Hence, it has the same dimensions as velocity, but it is not the speed at which the fluid moves. To derive the average fluid velocity in the pore space from the Darcy velocity, it is necessary to divide the latter by the effective porosity of the rock through which flow occurs (i.e., if the effective porosity of the rock is 10% of the total volume of the rock, the average fluid velocity in the pores is ten times the Darcy velocity). The volume of fluid flowing across the cross-section over a given time, which is the relevant quantity for determination of radionuclide flow through the seals (i.e., this is strictly not a "release", since there are additional barriers, such as upper levels of the shaft and the Culebra, between the seals and the accessible environment)—given a known radionuclide concentration in the fluid—is the same for both calculations, that is, with or without porosity.

Conclusion

These approximate calculations confirm the general impression that effective sealing of the repository is achievable using the designs proposed by DOE.

Appendix D

Creep Behavior of WIPP Salt

One of the main attributes of salt as a rock formation in which to isolate radioactive waste is the ability of the salt to creep, that is, to deform continuously over time. Excavations into which the waste-filled drums are placed will close the salt eventually, flowing around the drums and sealing them within the formation. A good understanding of the rate of closure and associated phenomena, such as the development and healing of fractures around the excavations, is essential for the design of an effective repository in salt.

The program of investigations into the mechanics of deformation of salt at the Waste Isolation Pilot Plant (WIPP) has been more extensive and comprehensive than any previous studies world-wide. Carried out over the past decade and a half, it has included theoretical studies coupled with small-scale laboratory tests and full-scale field investigations underground at WIPP.

Initial predictions of the rate of closure of underground excavations—based on deformation parameters derived from laboratory tests—were found to be some three to six times lower than the closure rates observed underground. Continued research to improve fundamental understanding has resulted in a very substantial reduction in this discrepancy. Thus, for the majority of the in situ cases studied, agreement between the closure rates predicted from small-scale laboratory test data and those actually observed is now within approximately 10 percent. Larger discrepancies can be attributed, in some cases, to exclusion of the following from the three-dimensional numerical codes used for the predictions:

• discrete slip surfaces (e.g., between the salt and anhydrite interbeds) in the vicinity of the excavations, and

• separation of interbeds and salt layers in the roof of the excavations.

These limitations are not serious for the assessment of closure performance of the excavations at WIPP since, in both cases, actual closure will somewhat exceed the predicted value. Thus, predictions will be conservative.

Although salt deformation appears to involve a complex interaction of a multiplicity of microscopic mechanisms, these combine to produce a relatively simple, essentially constant rate of room closure (Figures D.1 and D.2).

As shown in Figure D.3, the strain-time behavior of salt can be considered to consist of several regions, namely, ε_E, an elastic strain, which appears immediately upon loading (the amount of elastic strain increases with increase in the applied stress); this is followed by

1. a region of primary or transient creep strain;

2. a region of secondary or "steady-state creep" (where the strain rate $\dot{\varepsilon}_p$ is essentially constant); and

3. a region of tertiary or "accelerating creep" strain.

Accelerating strain indicates a progressive disintegration of the salt structure, leading eventually to collapse.

It is well known that the strain rates involved in salt creep increase in a highly nonlinear manner with increases in temperature and applied stress.

An extensive series of laboratory investigations has led to the definition of a new constitutive model (i.e., relationship between steady-state creep rate and applied load) for WIPP salt. This model is known as the modified Munson-Dawson or multimechanism deformation (M-D) constitutive model.

In the M-D model, the total steady-state creep rate ($\dot{\varepsilon}_s$) is considered to be the sum of three component rates, each dependent on a different fundamental mechanism to creep in the salt: Thus

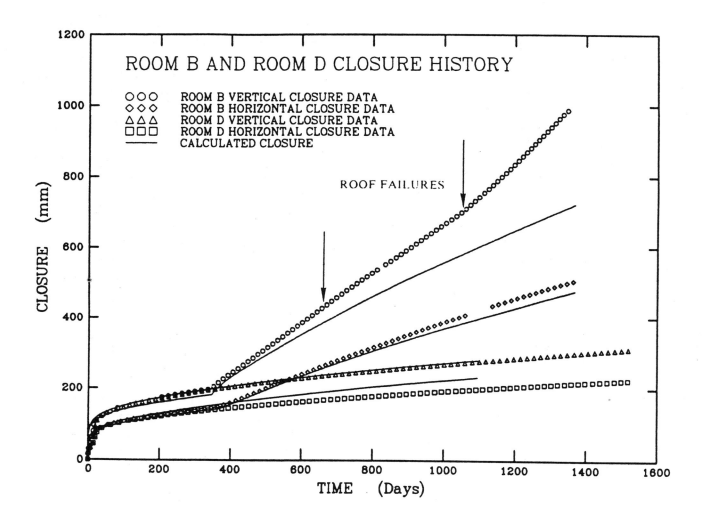

FIGURE D.1 Comparison between predicted and actual room closure rates for Rooms B and D. Note the increase in measured closure compared to calculated closure as roof failures develop. Source: Munson (1996).

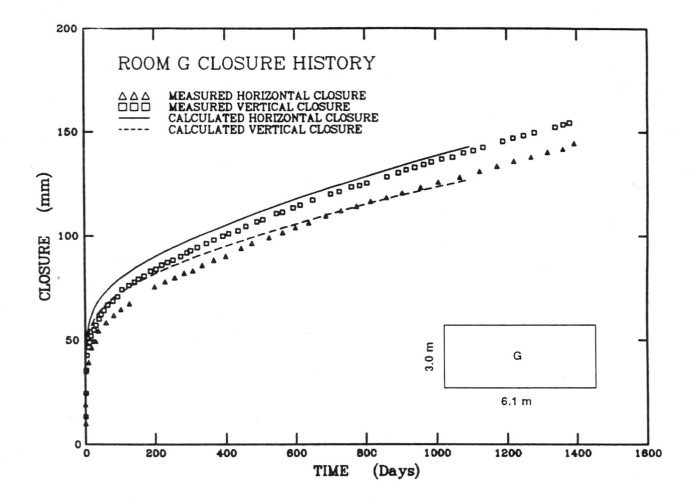

FIGURE D.2 Comparison between predicted and actual room closure rates for Room G. Source: Munson (1996).

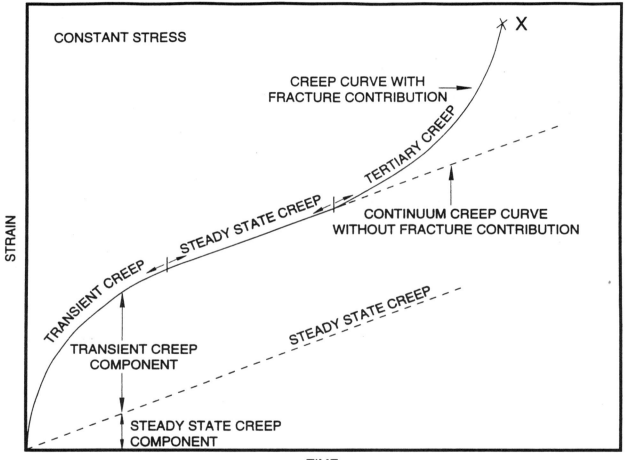

FIGURE D.3 Classical creep-deformation behavior of salt. Source: Munson et al. (1995).

$$\dot{\varepsilon}_s = \sum_{i=1}^{3} \dot{\varepsilon}_{s_i} \tag{1}$$

where the individual steady-state rates of the three relevant mechanisms are given by

$$\dot{\varepsilon}_{s_1} = A_1 e^{-Q_1/RT} \left(\frac{\sigma}{\mu}\right)^{n_1} \tag{2a}$$

$$\dot{\varepsilon}_{s_2} = A_2 e^{-Q_2/RT} \left(\frac{\sigma}{\mu}\right)^{n_2} \tag{2b}$$

$$\dot{\varepsilon}_{s_3} = H(\sigma - \sigma_0)\left[B_1 e^{-Q_1/RT} + B_2 e^{-Q_2/RT}\right]\sin h\left[\frac{q(\sigma - \sigma_0)}{\mu}\right] \tag{2c}$$

where A and B are constants; Q is the activation energy; T is the absolute temperature; R is the universal gas constant; μ is the shear modulus; σ is the generalized stress; n is the stress exponent; g is the stress constant; σ_0 is the lower stress limit of the dislocation slip mechanism; and H is the Heaviside step function with argument $\sigma - \sigma_0$ [i.e., $H(\sigma - \sigma_0) = 0$ for $\sigma < \sigma_0$; 1 for $\sigma > \sigma_0$].

A complete list of the values of the various constants in Equations (2a - c) for WIPP salt is presented in Munson (1996). For the current discussion, it is sufficient to note the following:

1. WIPP underground excavations will not be subject to any significant temperature changes (except for remote—handled transuranic waste locations, where some moderate temperature increases may occur). Thus, temperature dependence on creep closure need not be considered to a first approximation.

2. The value of σ_0 has been determined to be 20.57 MPa (megaPascals). For most of the underground locations at WIPP, since the initial stress state is isotropic (lithostatic pressure in every direction) at approximately 15 MPa, then $H(\sigma - \sigma_0) = 0$, implying that $\dot{\varepsilon}_s = 0$ for these locations.

3. Using the values $n_1 = 5.5$ and $n_2 = 5.0$ for the exponents in Equations (2a) and (2b), the steady-state strain rate $\dot{\varepsilon}_s$ reduces to the expression

$$\dot{\varepsilon}_s = k_1(\sigma)^{5.5} + k_2(\sigma)^{5.0} \tag{3}$$

where the generalized stress $\sigma = (\sigma_1 - \sigma_3)$; σ_1 and σ_3 are the maximum and minimum principal stresses, respectively; and k_1 and k_2 are constants representing combinations of terms in Equations (2a) and (2b), respectively.

For simplicity, it is assumed that both components of Equation (3) have the same exponent (i.e., assuming $n_2 \approx n_1 = 5.5$), so Equation (3) can be written as

$$\dot{\varepsilon}_s = k(\sigma_1 - \sigma_3)^{5.5} \tag{4}$$

where $k = k_1 + k_2$. It is seen that the stress dependence of the creep rate is very strong. For example, if an open borehole in salt closes at the steady-state rate $(\dot{\varepsilon}_p)_{open}$ at some depth h (meters) (i.e., where the lithostatic [driving] stress is $0.023h$ MPa), then a brine-filled borehole (with a hydrostatic pressure in the hole of $0.01h$ MPa) will close at a rate $(\dot{\varepsilon}_p)_{\text{fluid-filled}}$, where

$$\frac{\dot{\varepsilon}_{\text{fluid-filled}}}{\dot{\varepsilon}_{\text{open}}} = \left(\frac{0.023h - 0.01h}{0.023h}\right)^{5.5} \approx \frac{1}{37}$$

Thus, if an open unlined borehole closes in approximately 100-200 years, an unlined, fluid-filled borehole at the same depth would not close for 3,700-7,400 years. This is a significant change when estimating the consequences of human intrusion events.

Research at WIPP has also confirmed the following:

1. The stress at which inelastic flow of salt occurs is governed by a Tresca or maximum shear stress yield criterion, rather than the more frequently used Von Mises, or maximum octahedral shear stress yield criterion. Given the highly non-linear dependence of the strain and strain rate on stress level, the difference between the two criteria can be very significant. A part of the initial discrepancy between prediction and observation was corrected by making the change to a Tresca criterion.

2. The value of the shear stress at which inelastic strain develops tends toward very low levels at low strain rates. At the very low strain rates associated with geological loading, the yield stress of salt is essentially zero. In other words, at these extremely low loading rates, salt behaves as a (viscous) fluid, unable to maintain a shear stress. This is seen in equation (4) above, which indicates that the salt will have a nonzero strain rate $(\dot{\varepsilon}_s)$ as long as $\sigma_1 \neq \sigma_3$. Thus, over long times (especially geological time—here 200 million years) creep will continue until $\sigma_1 = \sigma_3$, (i.e. the in-situ stress state is isotropic).

Hydraulic fracturing measurements of in situ stress conditions of WIPP confirm that the stress state in the salt is isotropic. The isotropic condition and the creep flow characteristic of salt imply that salt in situ should be essentially impermeable, since the connected pathways needed to allow flow of fluid would lead to localized stress concentrations in the vicinity of the connected cracks, etc., that represent the permeability.

Such concentrations (stress differences) would produce flow of the salt, leading to closing of the pathway and elimination of the permeability.

3. At the higher strain rates associated with stress redistribution around excavations, salt can behave in a much more brittle manner. Fractures can and do develop progressively in the walls, roof, and floor of the excavation as the result of stress concentrations around the excavations. These fractures produce what is known as the disturbed rock zone (DRZ) around the excavations. The DRZ is important on several counts:

• It could provide a region of connected fractures that allow a high-permeability path around the excavations and brine access to the waste-filled drums. Such fractures (and DRZs) appear around excavations in other rock types, but salt has the attractive feature that fractures in the DRZ in salt can "heal" (i.e., disappear) with the buildup of a "back-stress" (i.e., resistance to closure generated by waste and/or backfill or a shaft lining or seal). This healing can be inhibited by the presence of water in the fractures as back-stress builds up. In the shaft seal designs discussed in Chapter 4, the DOE must take special measures to ensure that the fractures remain "drained" during the period of back-stress buildup. It is worth noting, however, that the fluid pressure required to resist crack closure can develop only if the fluid is sealed in the crack, that is, if it is not part of a permeable connected system of fluid flow.

• Since the DRZ develops in the floor of the excavations as well as the roof and sides, this zone serves as a depository and conduit for any brine flowing into the waste-filled rooms. Because closure of the waste-filled rooms involves floor "heave" as well as roof "sag," the waste drums will tend to be lifted above the brine, especially during the early years after loading of the rooms with waste. This phenomenon has important implications for the question of gas generation due to corrosion of steel waste drums at WIPP.

The practical implications of these deformation properties of salt are discussed in Chapters 3 and 4.

Appendix E

Actinide Source Term

As background for preparing its review of the actinide source term (AST), the AST subcommittee (Clark, Ewing, and Krauskopf) was supplied with abundant literature, including both in-house documents and articles from refereed journals that were considered by the Department of Energy (DOE) to be relevant to the Waste Isolation Pilot Plant (WIPP) program—many of which are listed at the end of Chapter 5. In addition, the subcommittee visited Los Alamos National Laboratory (LANL) in 1994 to view the proposed experiments with real and actinide-spiked waste (these are described as the Source Term Test Program [STTP]). In 1995 and early 1996, DOE scientists and contractors met with the AST subcommittee several times (March 6, April 27-28, and June 29, 1995; and February 14, 1996) to present the rationale for the AST program and some data from experiments completed in the past, and to describe experimental work in progress or still being planned (Novak, 1995a, b).

As late as April 1996, the subcommittee was still receiving revised versions of the proposed test plans, but no results of recent experiments or of the STTP program at LANL were available for review. The subcommittee was given a conceptual outline of the use of actinide source term data in the performance assessment (PA; Ruth Weiner, Sandia National Laboratories, personal communication to S. Clark, January 18, 1996); however, in the absence of experimental data explicitly used in the performance assessment, it was not possible to review this part of the program. Substantive review of AST results must await completion of at least a major part of the proposed work. *To the extent that the actinide source term is an important parameter in the performance assessment and the determination of compliance, it will be necessary to review carefully the final data package, the selection and use of conceptual models, and the final incorporation of data and models into the performance assessment.*

The remainder of this appendix discusses specific issues related to some of the tests planned as part of the DOE experimental program for the AST. The test plans may have changed after they were presented to the subcommittee and even before the experimental program was initiated; thus, these comments are applicable only to those portions of the test plans that have not changed.

The principal laboratories, principal investigators, and status of various components of the AST program are listed in Table E.1. This provides an indication of the materials that were reviewed by the AST subcommittee.

EXPERIMENTAL WORK ON ACTINIDE SOLUBILITIES

Pitzer Parameters

The application of Pitzer parameters (Pitzer, 1973, 1977) specifically to concentrated saline solutions has been described by Harvie et al. (1984) and by Felmy and Weare (1986). The methods of calculating solubilities suggested by these authors have been applied to the WIPP brines by DOE researchers. Particular advantages of using the Pitzer approach, rather than other means of correcting tabulated thermodynamic data to conditions of concentrated electrolytes, include the following:

1. The model is designed specifically for the geochemical conditions anticipated at WIPP and is well described in the peer-reviewed literature.

2. Data are available for many of the constituents in brine, making additional experiments unnecessary.

3. The model provides for complex interactions among more than just two species, reflecting the situation anticipated in highly saline media.

TABLE E.1 Actinide Solubility Work

Program Experimental Component	Principal Investigator (SNL/Subcontractor)	Subcontractor Laboratory	Date Initiated	Experimental Termination Date
Dissolved Concentration Model - Oxidation State Analogy				
An(III) Model Inorganic Interactions	Novak/Felmy, Rai, Rao	PNL	Jul-93	Apr-96
An(IV) Model - Inorganic Interactions	Novak/Felmy, Rai, Rao	PNL	Jul-93	Apr-96
An(V) Model-Part I Inorganic Interactions	Novak/Al Rafai	LBL	Feb-95	Mar-96
An(V) Model-Part II Inorganic Interactions	Novak/Runde	GTS/ITS	Feb-95	Jul-96
An(VI) Model- Inorganic Interactions	Novak/Palmer	LLNL	Jan-95	Mar-96
Organic Interactions An(III, IV, V, VI)	Novak/Moore	FSU	Feb-95	Apr-96
Alternative An(VI) Model-SIT Formalism	Novak/Jenecky	LANL	May-95	Mar-96
SNL Data Interpretation	Novak, Moore, Weiner	SNL	Feb-95	Feb-96
Dissolved Concentration Model - Challenges to the Oxidation State Model				
Tests of Conceptual Model Against Exper. with Actual TRU	Crafts/Villareal	LANL	Liter Scale: Mar-95 Drum Scale:	3-5 yrs. from initiation or steady state
Waste - STTP			Jun-95	
Empirical Solubility for An(IV, V, VI) in WIPP Brines	Moore/Reed	LANL	Jun-95	Sep-96
Further Oxidation State Studies	Weiner/Clark Felmy, Rai	LANL,PNL	Jun-95	?
Oxidation State Studies in STTP Experiments	Weiner/Clark	LANL	June-95	?
Mineral Fragments		SNL	Oct-93	Jan-96
Actinide Intrinsic Colloids		LLNL,LANL	Jul-95	May-96
Humic Substances		SNL, LANL, Gatson	Jul-95	Jul-96
Microbes		BNL, LANL	Jul-95	Feb-96

Source: Papenguth and Behl (1996 a,b); Novak (1995b); Novak et al. (1995).

An additional general cautionary note is in order regarding the estimation of solubilities in concentrated solutions with the use of the Pitzer parameters. The thermodynamic formalism involved in such estimates is based on the assumption that equilibrium exists between material in solution and solids in contact with the solution—in other words, that all reactions leading to formation or dissolution of solids go rapidly to completion. There is no assurance that these assumptions are correct for the experiments or for in situ WIPP conditions. In fact, some of the experimental work has shown that concentrations in a solution, after a solid has precipitated, may change slowly over the course of months or a few years as the solid alters its form, for example, from an amorphous precipitate to a crystalline phase (Nitsche et al., 1994).

Because changes in the solid are generally in the direction of greater and greater stability (Ostwald's step rule), and thus lower and lower solubility, such changes commonly are thought to ensure concentrations of actinides lower than those first measured, and hence not to be important in estimating cautiously conservative values for actinide solubilities. As a general rule this is true; however, for solutions as concentrated, and with compositions as complex as WIPP brines, exceptions can readily be imagined.

Resolving issues such as this will require extending some of the solubility experiments well beyond their scheduled termination dates. These experiments can be conducted at the bench scale under carefully controlled conditions, so that the fundamental aspects of actinide solubility in brine can be determined. Such experiments may be needed for interpretation of the more complex experiments performed as part of the STTP at LANL. These longer-term experiments should not delay compliance of the repository if all else is favorable, but they represent one example of the kinds of research efforts that are recommended to continue after wastes are first brought to the repository.

Concerns

The following are specific concerns about details of the planned experimental work on actinide solubility:

• Most aspects of the experimental program and assumptions in the performance assessment are based on thermodynamic models; however, reaction kinetics may be important in determining the actinide concentrations measured in experiments or modeled in the performance assessment.

• One question is the suitability of the specific-ion interaction theory (SIT), activity coefficient formalism, which has been used in the test plan for the Alternative +6 Actinide Model (Grenthe and Wanner, 1992). This approach for estimating concentrations of actinides in the +VI oxidation state is not scientifically defensible, because SIT theory is designed specifically to extrapolate to zero ionic strength; it is not applicable at ionic strengths higher than 3.5 molal; and it is typically used for applications involving a single dominant electrolyte, rather than the complex brines anticipated at WIPP.

• Another question regards the use of experiments planned in the STTP as direct challenges to the submodel of solubilities. The STTP involves measuring actinide concentrations in solutions formed by allowing waste containers to stand for long periods in contact with brines having compositions similar to those at WIPP; in other words, determining concentrations under laboratory conditions set up to duplicate, as nearly as possible, the conditions expected in the WIPP repository. Such experiments will certainly provide useful qualitative data on actinide concentrations, and this program is applauded. However, the assumption that experimental conditions can be made similar in all respects to repository conditions over time seems tenuous at best. Experiments that challenge the thermodynamic model will require not only careful solution analyses as a function of extended periods of time (more than a year), but the identification of the actinide-bearing solids that are in equilibrium with the solution. This will be difficult, because many of the solids are poorly characterized and may have fairly complex compositions.

• Although the oxidation state model (the assumption that the chemistry of a given oxidation state is similar for all of the actinides) is an appropriate beginning to a difficult problem, deviations from the oxidation state analogy are well known in natural and experimental systems. Substantial experimental verification will be needed to establish the limits of this analogy.

EXPERIMENTS ON COLLOIDS

Although actinide concentrations in WIPP brines are expected to be low, the possibility of transport via colloids must be considered one means by which actinides may reach the accessible environment (Kim, 1994). There are two issues: (1) whether colloids will form in a concentrated brine, and (2) whether these colloids can transport significant quantities of actinides effectively.

In designing their research effort, Papenguth and Behl (1996a, b) distinguish four kinds of colloidal particles:

1. mineral fragments, hard-sphere or hydrophobic particles that may contain actinides or have sorbed actinides on their surfaces;
2. actinide intrinsic colloids, particles consisting of actinide macromolecules that may mature on standing into mineral–fragment type colloids;
3. humic substances, hydrophilic or soft-sphere particles that readily sorb actinide ions from solution; and
4. bacteria, which may themselves be colloidal-sized particles and may contain actinides as part of their structure or sorbed on their surfaces.

All types of colloidal particles are relatively unstable in strong electrolyte solutions, and some of the experiments are designed to show what concentration of electrolyte ("critical coagulation concentration," or c.c.c.) is required to produce destabilization and precipitation.

The procedure is simple: small amounts of a simulated Salado-Castile brine are added to samples of colloid until coagulation is noted. Mineral fragments and mature actinide intrinsic colloids are expected to be the most sensitive to electrolytes (i.e., to have the lowest c.c.c.), whereas humic substances and bacteria will not precipitate as readily. For these latter two types, DOE plans more elaborate experiments to explore the effects of pH and the addition of specific ions and organic solutes.

The simple laboratory experiments on colloids will be supplemented with transport experiments designed to show the extent to which various kinds of colloids may be retarded, by precipitation or sorption, if the solution containing them travels through the dolomite of the Culebra Dolomite. Solutions carrying the colloids will be made up to simulate brine concentrations expected in the Culebra, and samples of the solutions will flow through both crushed dolomite and intact-core columnar dolomite. Transport experiments with bacteria as colloidal particles will be especially important, because microorganisms may serve either to increase radionuclide transport by sorbing abundant actinide ions or to decrease transport because the organisms carrying sorbed actinides are themselves strongly sorbed on rock surfaces (Marshal, 1976).

Again, to the extent that colloid formation and transport remain important issues in the performance assessment, the proposed experimental programs may well extend beyond their planned end dates. Long-term experiments on colloid stability and transport may be required even into the operational phase of the repository.

RETARDATION EXPERIMENTS

During transport, actinide concentrations in brine may be lowered by sorption onto rock and mineral surfaces. Such phenomena can be studied by batch experiments to determine K_d values, column experiments with actual rock units from the relevant geologic horizons, or field-scale tracer tests. The limitations of laboratory-scale tests are discussed briefly in Appendix F as part of the discussion of regional hydrology. The inherent weakness of the concept of a retardation factor determined in laboratory experiments, combined with highly uncertain and variable values for the retardation, requires careful analysis of the laboratory data and some field-scale verification. The proposed suite of DOE experiments has been planned carefully and should give a good indication of the effectiveness of sorption in controlling actinide concentrations; however, no results of the experiments completed to date were available.

Concerning the column experiments, the sorbing materials to be used are cylinders of Culebra Dolomite, 10-50 cm long, cut out of the bed with a drill bit and mounted vertically in metal containers. Various samples of dolomite will be chosen to ensure adequate representation of different permeabilities, different grain sizes, and different proportions of clay mineral impurities. It is emphasized that the column experiments (plus, hopefully, the field experiments) should be carried on long enough and with sufficient variation in conditions to ensure an adequate replica of the natural situation. In particular, the amount of sorption provided by the corrensite coating on parts of the Culebra dolomite requires careful study, a point on which the State of New Mexico Environmental Evaluation Group (EEG) has expressed considerable skepticism regarding the DOE approach.

Appendix F

Regional Hydrogeology

APPROACHES TO STUDYING REGIONAL AQUIFER SYSTEMS

In the broadest sense, a primary objective of studying a deep regional aquifer system is to improve understanding of the relevant processes and properties. Even after a given study has defined the properties of such a system, there will always remain much uncertainty. A unique definition of the system and its properties is simply not attainable, regardless of the amount of data collected. Data collection is very expensive in deep regional aquifer systems, and the costs of data collection and analysis must be weighed carefully against the benefits of additional data. There is no clear-cut point at which it can be determined objectively or scientifically that additional investments in data collection are either necessary or superfluous. Rather, a subjective assessment must be made on the basis of some type of consensus and in light of the significance of the problem, technical considerations, and economic constraints.

Although it is obvious that the properties of a regional aquifer must be defined with some accuracy in order to permit reliable predictions of ground–water flow and transport, it may be less clear that it is also necessary to define accurately the boundary conditions, recharge, and discharge in order to fully define and calibrate a model of the aquifer system. Studies of regional aquifer systems are most comprehensive when the study area extends to the natural geologic or hydrologic boundaries of the system and includes as much of the natural recharge and discharge areas of the system as possible. In reviewing the relevance of hydraulic analyses to contaminant transport problems, Reilly et al. (1987, p. 9) state that "although the details of the flow system in the immediate vicinity of the contaminant plume are of the utmost importance, an understanding of the regional ground-water flow system is required in order to understand the local system."

Because of the uncertainty and variability (heterogeneity) inherent in deep hydrogeologic systems, data collection in general, and drilling of new test wells in particular, often encounter unanticipated difficulties and frequently lead to some "surprises" in the definition or understanding of the properties of the system. Although experienced hydrogeologists expect this, such surprises unfortunately often lead to delays and tend to confound efforts to keep data collection and analysis to a strict, planned time line or budget.

GROUND-WATER MODELS

Ground-water simulation models help analysts improve their understanding of the subsurface system at the Waste Isolation Pilot Plant (WIPP) site faster than otherwise would be possible by integrating site-specific data with a theoretical understanding of the principles of flow and transport. The models offer a quantitative and objective means of testing hypotheses, evaluating the sensitivity of parameters, and guiding the data collection and experimentation process. Furthermore, if the models are based on reasonable and appropriate concepts, if they are adequately tested and calibrated, and if analysts have high confidence in their reliability, then the models can be used to predict how a particular system will respond to a particular stress or change in boundary conditions. However, even a well-calibrated model is still an approximation of the real system and incorporates many simplifying assumptions.

Sources of Error

In most model applications, conceptualization problems and uncertainty concerning the data are the most common sources of error. All three aspects of error are discussed below with reference to the performance assessment (PA) models for WIPP.

Conceptual Errors

Conceptual errors are derived from theoretical misconceptions about the basic processes, dimensionality, and/or boundary conditions that are incorporated in the model. Conceptual errors include neglecting relevant processes and representing inappropriate ones. Examples of such errors are applying a model based on Darcy's Law to materials or flow regimes for which Darcy's law is inappropriate, or using a two-dimensional model where significant flow or transport occurs in the third dimension.

That the PA models have a number of simplifying assumptions built into them is not, in itself, a problem, but rather reflects the very nature of models. Questions about ground-water models more typically center on whether the assumptions are too simple or too complex and whether they are appropriate for the system and problem at hand. However, there will never be a sure or precise way of measuring this. The reasonableness of the assumptions is perhaps best evaluated by a consensus of experts. If the conceptual models on which the PA models are based are, in fact, judged to be either too simple or inaccurate, then all of the complementary cumulative distribution functions (CCDFs) and risk assessments derived from the PA must be questioned.

The PA models start with the assumption that the primary release of contaminants under the human intrusion (HI) scenario is from leakage into the Culebra Dolomite. This assumption is part of the conceptual model. The risks and consequences of leakage into other formations have not been assessed or documented thoroughly. Release into the Culebra also is constrained by assumptions about the permeability of the material filling the borehole through which leakage occurs. However, in previous PA analyses, the assumed borehole permeability and assumed depths of impermeable plugs appear to be somewhat arbitrary.

Many of the concepts built into PA models may be based on preliminary judgments and assessments that are not well documented. For those ideas that may be critical but are based on minimal evidence, a range of alternative conceptual models should be evaluated in sufficient detail to document their consideration and the basis for their elimination. Examples of alternative conceptual models that should be evaluated further include (but are not limited to)

- contaminant leakage from a HI breach into formations shallower than the Culebra;
- flow and transport through the Culebra as part of a three-dimensional aquifer system;
- impact of changes in the flow field due to transient effects, climate change, or other factors; and
- internal "structure" of the Culebra, including the existence of vertical fracture zones and correlations between transmissivity and other parameters.

More detailed discussions of model testing, model calibration, predictive accuracy of models, and related issues are presented by Anderson and Woessner (1992), Bredehoeft and Konikow (1993), de Marsily et al. (1993), and Konikow and Bredehoeft (1993).

Numerical Errors

The partial differential equations describing ground–water flow and transport can be solved mathematically by using numerical methods. The continuous variables of the governing equations are replaced with discrete variables that are defined at grid blocks (or cells, elements, or nodes). Thus, the continuous differential equation that defines hydraulic head or solute concentration everywhere in the system is replaced by a finite number of algebraic equations that define the hydraulic head or concentration at specific points. This system of algebraic equations generally is solved by using matrix techniques.

However, numerical methods yield only approximate (rather than exact) solutions to the governing equation (or equations); they require discretization of space and time. The variable internal properties, boundaries, and stresses of the system are approximated within the discretized format. Because numerical methods do not yield exact solutions, *numerical errors* will exist in the solution because of truncation errors, round-off errors, and numerical dispersion. As long as the sources of numerical errors are recognized and the magnitude of the error is controlled, the solution may be sufficiently accurate for use.

The solute-transport equation is in general more difficult to solve accurately using numerical methods than is the ground–water flow equation, largely because the mathematical properties of the transport equation vary depending on which terms in the equation are dominant in a particular situation. Thus, numerical errors often are negligible in solutions to the flow equation but can be significant in numerical solutions to the solute-transport equation.

Where solute transport is dominated by advective transport, as is common in many field problems, the transport equation approximates a hyperbolic type of equation (similar to those describing the propagation of a wave or of a shock front). In contrast, where a system is dominated by dispersive fluxes, such as might occur where fluid velocities are relatively low and aquifer dispersivities are relatively high, the transport equation becomes more parabolic in nature (similar to the transient ground–water flow equation).

In solving advection-dominated transport problems, in which a relatively sharp front (or steep concentration gradient) is moving through a system, it is numerically difficult to preserve the sharpness of the front. Obviously, if the width of the front is narrower than the node spacing, it is inherently impossible to calculate the correct values of concentration in the vicinity of the sharp front without using complex nonlinear mathematical functions or complex nonstandard numerical algorithms. However, even in situations where a front is less sharp, the numerical solution technique can calculate a greater dispersive flux than would occur by physical dispersion alone or would be indicated by an exact solution of the governing equation.

That part of the calculated dispersion introduced solely by the numerical solution algorithm is called *numerical dispersion.* The consequence of numerical dispersion is that the solute is artificially diluted, peak concentrations are lower than they should be, and the solute is distributed throughout a larger than actual area and volume of the aquifer. Although this alone may or may not represent a significant problem, the errors can in fact be magnified and accumulate if the model incorporates a representation of solute retardation processes in which rates or magnitudes are functions of the solute distribution and/or concentration gradients. Because the effectiveness of matrix diffusion and chemical retardation, for example, depends in part on the surface area of aquifer solids that are in contact with ground water containing the solute (or contaminant), numerical dispersion may yield erroneously high rates of matrix diffusion and chemical retardation. Therefore, special care must be taken to assess and minimize numerical errors that artificially would add "numerical" spreading to the calculated spatial distribution of a contaminant.

Data Uncertainty

Data errors arise from uncertainties and inadequacies in the input data. A reasonable way to accommodate such uncertainty is to estimate the mean and the uncertainty in each relevant parameter on the basis of available data, and then use the simulation models to make predictions based on *ranges* of values of the parameters, rather than on single values. Outcomes from the deterministic models can then be presented and assessed in a probabilistic framework. In essence, this is the approach taken by the PA models.

The reliability of site-specific ground-water models typically requires that input data be based on field measurements of critical parameters at a scale appropriate to the size of the model. Unfortunately, very few field measurements are available for the critical parameters describing the nature of fractures and retardation mechanisms in the Culebra.

Although the probabilistic approach is beneficial and appropriate, it must be cautioned that the results will be affected by assumptions about the statistical distributions and independence of parameters; that is, there is uncertainty in the estimates of uncertainty. The sensitivity of results to assumed frequency distributions should be evaluated.

In addition, the system behavior for some processes is governed essentially by the magnitude of nondimensional parameter groups (or ratios). The rationale for independently sampling the individual parameters comprising such a group is unclear and may result in unlikely parameter ratios.

Hydrologic Performance Assessment Models

Ground-Water Flow

In developing PA models for flow and transport through the Culebra Dolomite, estimates of recharge and transmissivity are keyed into the a priori assumptions that ground-water flow is two dimensional, steady state, and horizontal. In reality, the Culebra is one part of a three-dimensional flow system. Although flow may be predominantly horizontal, some vertical components of flow and transport undoubtedly exist, and they may significantly affect the flow, transport, travel times, and concentration distributions in the Culebra in areas downgradient from a potential release.

Fractures are the high-permeability avenues (or pathways) through which the fastest flow and transport of dissolved contaminants would occur. Fractures thus represent the most likely fatal flaw in site integrity if the HI scenario enables significant leakage of contaminants into the Culebra. In general, where vertical or near-vertical fractures exist, they

1. tend to persist across formation boundaries,
2. often yield some subtle but detectable expression or indication at the land surface, and

3. tend to "channelize" ground-water flow.

Because the presence of a few high-permeability fractures will have a large effect on solute transport, additional efforts to characterize fracture properties are warranted.

Confidence in PA Analysis of Regional Ground-Water Flow

The existence and importance of heterogeneity in the Culebra Dolomite is well recognized. Modern, sophisticated, and complex geostatistical methods have been applied to analyze the statistical properties of the variability in transmissivity and to estimate values of transmissivity in areas where no measurements are available. Accomplished with the assistance and advice of a team of world-class experts, this is a state-of-the-art approach to estimating point values and spatially correlated trends in transmissivity based on existing data, as well as to assessing the reliability of these estimates.

In 1995, a new set of wells was drilled into the Culebra for the purpose of conducting multiwell tracer tests. The site (designated as H-19) was selected in part because it is in the area where the coupling of geostatistical methods and ground-water modeling indicates a relatively high transmissivity zone. If hydraulic testing at this new site confirms prior estimates, it would provide a basis for increased confidence in previous analyses. This new drilling site also would provide a control area in which to apply and test the effectiveness of high-resolution geophysical methods and topographic analyses, in order to better define fracture zones. The result of such analyses could then be correlated with the results of planned tracer tests.

Some concern remains about scale effects on permeability and transmissivity. For example, a comparison of transmissivity values for wells H-14, H-15, and H-17 that were estimated from single-well tests (Beauheim, 1987) with values for the same wells as calculated from data collected during a larger-scale pumping test at H-11 (Beauheim, 1989) shows that the latter are 20 to 60 times greater. Garven (1995) states that fractures in carbonate aquifers tend to increase the hydraulic conductivity at the regional level. He presents a curve indicating that at a regional scale of 1 km or greater, hydraulic conductivity may be a few orders of magnitude higher than estimated from borehole measurements at scales of less than 100 m.

This is relevant to WIPP studies and of some concern to the committee because the assumed transmissivity distributions are keyed to transmissivity values measured in boreholes (see LaVenue et al., 1995, for a detailed discussion of the methodology and its application to the WIPP site). These types of borehole measurements typically sense aquifer properties only in the immediate vicinity (perhaps a few meters) of the borehole. If some of those borehole measurements underestimate the larger-scale effective transmissivity in that area, the regional transmissivity values generated may be too low for purposes of large-scale transport predictions. Although the generated transmissivity fields are calibrated by using observed head measurements from both the assumed steady-state flow field and transient hydraulic tests at a few sites, the heads are less sensitive to transmissivity variations than are seepage velocities.

Geostatistical analysis does not explain why transmissivity varies the way it does. It does not answer the question of whether higher transmissivities are related to increases in fracture density (decrease in fracture spacing), increases in fracture aperture, increases in fracture length (and connectivity), changes in fracture orientation, increases in primary porosity, changes in the amount of evaporite minerals dissolved, and/or changes in mineralogy.

Some might ask whether it is important or necessary to understand the cause of these variations in order to predict ground–water flow and transport. Finding the answers to such questions is not merely an academic exercise; the answers are relevant to the PA models and analyses because whatever is causing the changes in transmissivity will most likely also affect the effective porosity and retardation processes. Developing a better understanding of the causes

and controls of transmissivity variations may require improved geological and geophysical mapping of primary and secondary geological properties and characteristics of the Culebra dolomite, particularly fractures.

Franke et al. (1987, p. 2) state that "selection of the boundary surface and boundary conditions is probably the most critical step in conceptualizing and developing a model of a ground-water system." Neither the PA ground-water model of the Culebra nor any of the preceding deterministic models of the system extended to the natural boundaries of the regional aquifer. This makes it difficult to assess the reasonableness of the boundary conditions specified at the outer edges of the PA model, which encompasses only a small part of the entire regional system. It is similarly difficult to calibrate the flux through the Culebra and to distinguish between recharge-discharge in outcrop areas and recharge-discharge from leakage through confining beds. In addition to providing another example of a less than desirable level of understanding of the system of interest, this difficulty also casts doubt on the ability to evaluate and predict the impact of climate change because it is not known how a projected change in climate might interact with the aquifer system in general or with the Culebra in particular.

Future climate change may significantly affect the regional ground–water flow system. The PA assumes that the maximum amount of increased recharge to the Culebra can be represented by increasing assumed boundary heads in the model to the elevation of the land surface. However, because it is well known that heads in confined aquifers can be significantly higher than land surface, it is not clear that this assumption really represents a bounding case. If future climate change induces greater diffuse recharge to the water table, then the saturated thickness of the Dewey Lake Red Beds may increase significantly, resulting in greater transmissivity values and higher flow velocities. Since the recharge would likely retain a relatively low content of dissolved solids, use of the Dewey Lake as a drinking water source might increase significantly in the future. In such circumstances, the Dewey Lake Red Beds could prove to be a greater long-term environmental risk than the Culebra.

Solute Transport

Solute-transport processes include advection, in which solutes move with the average seepage velocity of the flowing ground water, and hydrodynamic dispersion, which has the net effect of causing the solute distribution to spread about the positions indicated by the mean seepage velocity. Of these two primary processes, advection is the most important to predict accurately, because it controls when the center of mass of a contaminant plume will arrive at the regulatory boundary. Any errors in the solution to the flow equation or in the estimates of effective porosity will affect advection.

The dispersion process in ground-water systems is still the subject of much active research. Although conventional theory holds that the dispersivity coefficients are generally an intrinsic property of the aquifer, in practice they are found to be dependent on and proportional to the scale of the measurement. Recent controlled field experiments and theoretical advances in the study of transport processes indicate that large-scale (macroscopic) dispersion results largely from spatial variations in velocity, caused by spatial variations in hydraulic conductivity and effective porosity (see Gelhar et al., 1992). In that sense, dispersion and advection are interrelated.

Experimental field studies of solute transport at sites that include a dense network of observation wells tend to show that contaminant plumes emanating from known point sources are narrower than expected and that actual dispersion is much less than anticipated by conventional theory (e.g., see Garabedian et al., 1991). The implication for a release of contaminants into the Culebra Dolomite under the HI scenario is that the resulting plume may be longer, narrower, and more sharply bounded than predicted with existing PA models. Overestimating dispersion would have the net effect of diluting the contaminant by spreading its mass through a larger volume of aquifer. In a homogeneous single-porosity medium for a nonreactive solute, this would have no net effect on the center of mass. However, in a heterogeneous system such as the Culebra, overestimating dispersion also forces too much solute into lower-velocity parts of the flow field. This will bias the results toward yielding lower releases during a given time period. Also, if the solute is subject to retardation, overestimating dispersion can lead to a bias of overestimating the magnitude of retardation, as discussed below.

Confidence in PA Analysis of Solute Transport

Because solute-transport processes could be critical to establishing a sufficiently low level of radionuclide releases at WIPP, achieving high confidence in the analyses is important. Although several tracer tests run in the field were valuable steps in this direction, they had to be run under forced-gradient conditions because of time and cost constraints. Running a long-term natural-gradient tracer test should be considered as well, because it would serve as an indicator of true long-term migration paths and velocities under conditions expected after a breach via the HI scenario. However, because such a test might require decades of monitoring, which could be done at very long intervals, it is not recommended that this type of test be performed prior to a compliance decision. Rather, it is suggested that its feasibility be evaluated for possible use as an operational and postclosure monitoring procedure.

Retardation Mechanisms

As solutes migrate through the aquifer, the apparent mean velocity of a particular solute may be retarded relative to the apparent velocity of the water because of several processes that may cause molecules or ions of that particular chemical species to be removed temporarily from the active flow field. Retardation processes thus provide another margin of confidence for the WIPP repository in that retardation will delay (but not prevent) the arrival of a contaminant at the regulatory boundary.

Matrix Diffusion and Physical Retardation

The PA models assume that solute transport through the Culebra Dolomite occurs through continuous sets of horizontal parallel fractures. However, some of the field tests have estimated parameter values for different types of fracture sets and distributions, such as orthogonal sets. The differences in conceptualization must not yield inconsistencies in the effective surface area of fractures per unit volume of aquifer between the field estimates and the PA models, because such inconsistencies might affect the calculated rates of matrix diffusion.

Overestimates of the amount of plume spreading can occur if the assumed dispersion coefficients are too large or if assumed patterns of heterogeneity artificially induce divergence of flow. These effects would be of minor consequence for transport of nonreactive solutes in a single-porosity system. However, if matrix diffusion is a simultaneous process, then the problem is more serious, since the net result would be too much retardation.

Inaccurate or imprecise initial conditions also can lead to significant artificial spreading of the contaminant plume. The HI scenario assumes that the contaminant is introduced into the Culebra by leakage from a borehole (or well) that has a diameter on the order of 0.3 m. In the 1992 PA model, that contaminant source is introduced as an initial condition at the finite-difference cell representing the well. Because that cell has dimensions of 125 m by 125 m, the initial width of the plume at the start of the simulation is perhaps several tens of times greater than it would actually be. The width of the plume then spreads out even more as it is advected and dispersed in a downgradient direction by the regional ground–water flow system.

If the initial width of the plume is much greater than it should be, it has the net effect of giving matrix diffusion an unrealistic "head start" and a growing "advantage" in retarding the movement of the contaminant plume by allowing the plume to be in contact with an erroneously large surface area of the matrix. In a fracture-dominated carbonate

aquifer, it is possible that contaminant plumes can migrate long distances with very little lateral spreading, which would minimize any retardation induced by matrix diffusion.

Numerical dispersion also causes artificial plume spreading. The SECO transport code solves the transport equation by using a type of finite-difference scheme. When finite-difference methods are used to solve advection-dominated transport problems, some amount of numerical dispersion is likely to be induced in the solution. Because the numerical code has not been widely available for outside testing and evaluation, it is difficult to know how much numerical dispersion is present. However, any numerical dispersion will necessarily result in overestimates of matrix diffusion.

Confidence in PA Analysis of Matrix Diffusion

The PA models assume that matrix diffusion will be controlled by the intergranular properties of the matrix (including porosity and tortuosity), the free water diffusion coefficient for a particular constituent, and the concentration gradient. *Tortuosity* is a parameter reflecting the fact that the actual tortuous flowpath that a fluid particle would travel through the liquid phase of a saturated porous medium is longer than the length of the porous medium sample. However, tortuosity has been estimated only indirectly in laboratory measurements on a limited number of Culebra samples (Kelley and Saulnier, 1990), where it is shown that the diffusion porosity is usually about two to eight times lower than the porosity measured by other standard methods on the same samples. It is unclear whether these laboratory measurements can be transferred to the field environment.

Field tests by the WIPP project to date have yielded somewhat ambiguous results regarding the occurrence and rates of matrix diffusion. If matrix diffusion must be relied upon to demonstrate compliance, further field studies are necessary to support the model assumptions. Field evidence is needed over moderate distance and time scales and could likely be obtained as part of a series of multiwell tracer tests. Sandia has proposed using single-well injection-recovery tracer tests to demonstrate the existence of matrix diffusion. However, it is doubtful that such tests could provide either unambiguous results or meaningful measurements of the relevant parameters.

It is also possible that in the field, solute transfer between the fractures and the matrix may be hindered by some type of skin effect or discontinuity at the interface between a fracture and the matrix. Sharp et al. (1995) note that fracture skins are an important and integral part of natural, fractured rock systems and that skin permeabilities and mean pore diameters may be reduced by more than an order of magnitude relative to the matrix. They state that "an extreme case is where the fracture skins have zero permeability and porosity so that no interchange of fluids, colloids and solutes between the fracture and the matrix is possible." Analytical methods are available to interpret the results of tracer tests conducted in fractured aquifers having skin effects (see Moench, 1984; Sharp et al., 1995). The presence of a skin effect would render meaningless any projections of matrix diffusion rates based solely on the physical properties of the matrix. This possibility needs to be given further consideration.

Reactions and Chemical Retardation

As noted in Chapter 6, the appropriateness of predicting the transport of reactive solutes on the basis of a retardation factor (R_f) is questionable. Furthermore, if the use of a retardation factor represents a poor conceptual model, then it is doubtful that the theoretical difficulties can be overcome by simply evaluating a range of values for R_f. For example, if the governing reaction is chemical precipitation of a solid phase, rather than sorption, the species either is less likely to redissolve into the liquid phase in which transport occurs or will do so at a much slower rate. In this case, use of a retardation factor will probably indicate faster and further transport than might occur (but it will still depend on the precipitation rate).

Alternatively, too much retardation could be calculated if the governing process is indeed a type of sorption controlled by surface chemistry. For example, sorption is sometimes characterized by a Langmuir isotherm, which is one type of nonlinear isotherm (although it should not be inferred that a Langmuir isotherm would necessarily be applicable at the WIPP site). In such cases, retardation is greatest at very low concentrations, but at high concentrations the slope of the isotherm approaches zero and the medium will not adsorb additional solute (see Figure F.1). Thus, if sorption is actually governed by a Langmuir isotherm or a similar nonlinear isotherm, applying a retardation factor may lead to a calculation of too much retardation at the high end of the concentration range—the very part of the spectrum at which the greatest toxicity is likely to exist. This illustrates a major weakness of the linear R_f model: it places no upper limit on the amount of the chemical that can be sorbed. However, if it is known that the solute concentration will always remain relatively low, then even if the governing isotherm is nonlinear over the full range of possible concentrations, the sorption behavior within the range of expected conditions will not differ materially from that predicted by a linear isotherm model. If chemical retardation is to be relied on for compliance of WIPP, it must first be demonstrated that the use of a retardation factor is an appropriate representation of reaction processes in the Culebra for the solutes of interest over the transport scales anticipated.

Although most types of reactions tend to retard the transport of a reactive solute, the presence of colloids in the fluid may increase the mobility of solutes that otherwise are retarded or relatively immobile. For example, Ibaraki and Sudicky (1995) cite a case at a liquid waste disposal site at Los Alamos, New Mexico, where both plutonium and americium migrated more than 30 m. They report that predictions based on laboratory results that ignored the role of colloids severely underestimated the migration as just a few millimeters. Colloid-facilitated transport is an issue that warrants additional consideration to justify its exclusion from the PA evaluation.

Confidence in PA Analysis of Chemical Retardation

Not only is the concept of a retardation factor a theoretically weak one for predicting radionuclide transport through the Culebra, but the underlying parameter values for the equivalent K_d are highly uncertain, spatially variable, and based on minimal laboratory experiments that have questionable relevance to the large-scale field environment. Most of the available K_d values obtained by the WIPP project were determined from batch experiments on crushed or powdered samples of rock. There is little evidence that these K_d values are transferable to the field environment, in part because the effective surface areas for water-mineral interactions in crushed samples would be very different from those in situ. For example, Lienert et al. (1994) studied uranium transport in a sand and gravel aquifer and found that the in-situ K_d is much lower than most K_d values obtained from laboratory experiments.

Significant spatial variability in values of K_d may exist and, if so, is probably related to variations in mineralogy. However, in contrast to the consideration of spatial variability in transmissivity, no data are currently available to provide a basis for estimating the magnitude of such variability. A recently published review of transport of reactive contaminants in heterogeneous media concludes that "the transport of reactive solutes appears to be influenced most greatly by subsurface heterogeneity and rate-limited mass transfer interactions" (Brusseau, 1994, p. 306). In discussing the application of theory and laboratory measurements to field-scale systems, the paper further concludes (p. 307): "Of course, true field-scale behavior can be obtained only by performing experiments in the field."

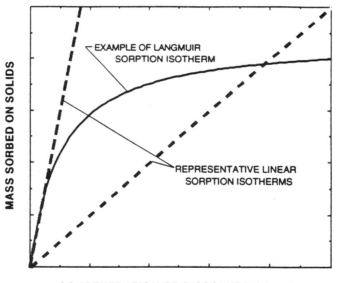

CONCENTRATION OF DISSOLVED SOLUTE

FIGURE F.1.   Schematic comparison of representative linear isotherms, describing instantaneous reversible sorption-desorption reactions, with an example of a nonlinear Langmuir isotherm. The distribution coefficient K_d is equal to the slope of the linear isotherm and is directly proportional to the solute retardation factor R_f. Source: Modified from Goode and Konikow (1989).

Appendix G

An Overview of WIPP Compliance Issues

This appendix is a nontechnical synopsis of the major themes and findings of this report. The primary goal of this report is to present a scientific and technical appraisal of DOE activities in connection with the Waste Isolation Pilot Plant (WIPP) radioactive waste isolation program. This entails detailed discussions of complex issues related to engineering measures, biochemical processes, and hydrogeological aspects of the site. Radioactive waste isolation is also a topic of considerable interest and concern to the general public, and it is important that the public understand the basic concepts underlying the design and operation of the WIPP facility.

This appendix aims to describe the WIPP project clearly and concisely by using a "nonspecialist" approach—that is, one that avoids specialized technical terminology as much as possible—to discuss salient issues pertaining to the project. It traces the history of the facility and summarizes the committee's main concerns, conclusions, and recommendations regarding Department of Energy (DOE) activities at WIPP.

DEEP GEOLOGIC DISPOSAL

The objective of the nation's nuclear waste disposal program is to place solid waste in a location where it cannot return to the biosphere by any foreseeable natural process. A generic method has been selected: bury the material deep within the earth's rocky crust. Salt beds are particularly suitable because they are very stable over geological time scales and have the ability to flow and permanently seal the excavations in which the waste is placed. WIPP is located in such a formation.

The plan for waste disposal at WIPP is to excavate a series of underground rooms in the salt deep below the earth's surface. Radioactive waste will be emplaced in the rooms, and then the shafts to the surface will be filled with salt and other materials to close the facility.

The weight of the overlying rocks will compress the disturbed salt and reduce its ability to transmit fluids and gases to the very low values of the undisturbed salt, thereby "sealing" the repository.

The worldwide use of salt cavities to store compressed gases attests to the remarkably low permeability of deeply buried salt (Berest and Brouard, 1996). Salt also has a well-known ability to flow under applied stresses. At the depth selected for the repository, crushed salt will reconsolidate to the same low permeability as that of the original formation, and even large cavities eventually will disappear.

Numerical modeling can predict the flow behavior of salt far into the future for any defined situation. The flow is slow enough to allow ample time for underground operations, yet fast enough to be essentially complete in a hundred years or so.

PROJECT ADMINISTRATION AND REGULATION

The WIPP project was begun in the mid-1970s. It has been the responsibility of DOE and its predecessor agencies, but many years passed before it was determined that the U.S. Environmental Protection Agency (EPA) would be responsible for certification. Disposal regulations then had to be developed in parallel with field activities. The "Final Rule" for the Criteria for Certification was published on February 9, 1996, as this report was being prepared.

CERTIFICATION CRITERIA

The long timeline associated with transuranic (TRU) waste poses problems for the regulation of WIPP, just as it does for design and construction. It seems obvious that protection of human health should be fundamental, yet it is impossible to predict what humanity will be doing in the vicinity of WIPP during the next ten millennia. Given the nature of the EPA standard and the apparent lack of release scenarios for WIPP under undisturbed conditions, it is not necessary to calculate individual doses as part of a compliance analysis. The part of the EPA standard that is relevant to WIPP is a release limit that avoids the necessity of making arbitrary assumptions about population distributions and human activities in the far future. Unfortunately, this release limit fails to account for one of the advantages of the WIPP location—namely, that most of the ground water in the area is so salty (i.e., brine) that it is undrinkable.

The release limit is specified in a rather surprising way: not as an absolute amount, but as a fraction of the waste inventory. The amount is measured conventionally in curies (a measure of radioactivity) rather than mass. This specification can be translated approximately for nonspecialists as follows: the expected waste inventory is of the order of 6 million curies, and the permitted release fraction of plutonium-239, the dominant radioactive isotope of concern, is 0.0001 of that, or 600 curies. This is the same as a mass of Pu-239 of about 20 pounds.

Regulation in terms of release did not entirely relieve EPA from the necessity of making arbitrary decisions. How long a period is implied by the word "permanent"? EPA settled on 10,000 years, although the half-life of plutonium is 24,000 years. How long can one assume "administrative control" of the WIPP area after closure? The EPA chose 100 years. What kind and frequency of human intrusion should be considered? The EPA chose, reconsidered, then chose again. DOE is now required to assume that drilling will continue in the same manner and intensity as today for the next 10,000 years. What credit should be given to the present generation for assuming extra risks that reduce risks to future generations? The EPA is silent on this point.

The simultaneous presence of radioactive and chemically toxic wastes in WIPP presents a "mixed-waste" conundrum. Must EPA enforce Resource Conservation and Recovery Act (RCRA) provisions (relating to disposal of hazardous waste) at WIPP? A recent law (P.L. 104-201) removes this requirement at the federal level, providing DOE with an exemption from the need to demonstrate that volatile organic compounds (VOCs) would not escape from the repository. The justification for this exemption is based on the assessment that the toxicity of the radionuclides in WIPP is far more important than that of the chemicals.

CERTIFICATION OF WIPP

In October 1996, DOE intends to submit to EPA an application for a certificate of compliance. A critical item in the documentation package will be the results of the performance assessment (PA) group at Sandia National Laboratories (SNL). PA is a set of methodologies used to model the long-term behavior of radioactive waste disposal systems.

The purpose of a performance assessment is to answer three questions:

1. What can go wrong?
2. How likely is it to go wrong?
3. What are the consequences?

The first question was answered years ago by inviting interested parties to submit "scenarios" in which human or natural events lead to the release of radionuclides from the repository. After extensive review of a large number of proposed scenarios, eight were selected on the basis of their credibility and significance.

The PA group at SNL is charged with answering the other two questions, using an approach mandated by EPA. The group is developing a large computer model of WIPP. The model interconnects many large special-purpose programs, addressing, for example, site hydrology, geology, and design features. Into this model go descriptions of the eight scenarios, along with data on the many physical and chemical parameters and rate constants required for the computations.

Providing data for this model has been and continues to be a major activity of SNL and its subcontractors. Laboratory and field work play a large part; other approaches include the systematic "elicitation" of expert opinion for matters that are difficult to measure (e.g., the solubility of certain actinides in saturated brine) or impossible to measure (e.g., the best way to communicate to future generations the danger of digging within the WIPP reservation).

In one of the eight scenarios, the repository remains undisturbed by humans after it is closed. For this "undisturbed" case, neither the committee nor DOE's contractors could think of any way in which there is a credible possibility for release of radionuclides.

The other seven scenarios involve human intrusions (HI) of three kinds, alone and in combination, and with various frequencies. A variety of possible circumstances, such as the presence or absence of brine, accompanies each of these "disturbed" scenarios. The existence of many possible scenarios and many kinds of data makes the computations very complex. The computer can handle this complexity, but the computations are nearly inscrutable to all but specialists.

A problem arises for certain data used in the computations: what numerical value should be used when only a broad range of possible values is available? Choosing the "worst case" value (i.e., the least desirable value) of every range maximizes conservatism but introduces an unnatural bias. When faced with such uncertainty, statistical theory provides a means to make predictions on the basis of a defined range in values. This is accomplished by making many simulations for the same scenario, but in each simulation a different value for each uncertain parameter is selected randomly. If implemented carefully and if a large number of such model simulations are made, the average of the numerous predictions can provide a reasonable estimate of the likely outcome for comparison with the compliance criteria.

The output of such simulations is a calculation of hypothetical releases of radionuclides from the repository. Associated with each calculated release is a calculated likelihood (probability) that such a release will occur. Each of these calculations of release and probability of release is called a "realization." The realizations are thus a series of related probabilities and releases that can be plotted on orthogonal axes (e.g., horizontal and vertical, with releases on the horizontal axis and probability on the vertical axis). The releases appear as points, and the curve connecting them would look like a stepped line or a smooth curve, depending on the number of realizations. The analysis of each scenario includes many realizations.

Finally (see Figure 2.1), scenario probabilities are combined for graphical display as complementary cumulative distribution functions (CCDFs). Plotted on the same scale is a single, rather simple graph representing in probability form the EPA maximum-release specifications.

CHEMISTRY, BIOLOGY, AND GEOTECHNOLOGY

It has been the practice of the PA group to be conservative in modeling. This is sensible, except that two or three excessively cautious estimates may combine to produce an unrealistic model result, or a conservative assumption may come to be regarded as a "given." Therefore, it is important that conservative assumptions be confronted constantly with new information as it is developed. This confrontation has not always occurred in the WIPP PA.

For example, the pre-1996 PA efforts used release scenarios that assumed the following:

- all storage rooms were interconnected and able to be flooded with brine;
- brine might dissolve all the steel containers and all the TRU elements within them; and
- gas production by chemical and biological processes might raise the pressure within the sealed repository to and beyond the pressure of the overburden (lithostatic pressure).[1]

These assumptions may be erroneous. The amount of brine that can enter the repository from rock in the immediate vicinity is in fact quite small in the undisturbed case. Brine occurs in the Salado Formation as pockets in "seams" of clay and/or anhydrite (calcium sulfate), and in porous regions that occur irregularly within the otherwise nonporous salt. There appears to have been little or no interconnection within the pockets of brine in the salt for millions of years.

---

[1]Gas generation is of particular concern because, if excessive, it could lead to a rise in pressure capable of rupturing the repository, with generation of cracks that would provide a pathway for brine to enter and radionuclide-laden brine to migrate out.

Excavation disturbs the brine pockets in the vicinity of the cavity. During excavations, the rock immediately around the excavation becomes fractured, leading to what is called the "disturbed rock zone" (DRZ). Some of the fractures penetrate the brine pockets and cause the brine to drain toward the excavation. The rate of brine collection has been observed to peak within a few years and then drop off as the DRZ ceases to grow.

Hydrogen generation from reaction between the brine and the waste containers also has been overestimated. Laboratory work has established that brine dissolves steel only when there is actual contact between liquid and metal; where there is no contact, only a thin "tarnish" film forms.

Microbiological experiments have not yet given a definitive answer about biological gas generation, but this too requires liquid water. Moreover, there are fundamental reasons for doubting that organisms can continue to be active metabolically for long periods in an isolated environment containing only limited amounts of the many organic and inorganic factors that are required, even when the food (carbohydrate) supply for microorganisms is unlimited.

The committee uses these considerations to judge that there is only a small likelihood that significant amounts of brine will enter the repository and react with the waste to produce hydrogen gas.

On the other hand, brine flooding does occur in scenarios in which a borehole connects the repository with a pressurized brine reservoir in the Castile Formation below the Salado. The possibility of this type of event is given serious consideration, in part because such a reservoir was encountered in a test boring in the 1970s.

THE ACTINIDE SOURCE TERM

Scenarios that assume brine flooding followed by leakage of contaminated brine upward into a near-surface aquifer require data on the possible amount and concentration of radionuclides. The "actinide source term" (AST) refers to the kinds and amounts of dissolved or suspended radionuclides that could be released to the biosphere.

The total quantity of radioactive materials expected to be emplaced in WIPP has been estimated from inventory records of the sites where wastes are currently stored. However, estimates of the proportion of waste that actually would be dissolved in the case of brine flooding are very uncertain. This seems like a straightforward chemistry problem, but few laboratories are equipped to handle plutonium (a large component of the radionuclide inventory at WIPP). There is little useful information about this issue in the published literature.

An experimental study of this problem was begun at Los Alamos National Laboratory (LANL) in 1994, but no data have yet become available. The current suite of experiments is scheduled for completion in May 1996. It is possible that useful data will be available in time to be included in the PA in the October 1996 certification package. Evidence from the German waste isolation program, which is also considering a salt repository, suggests that plutonium solubility in brine is likely to be low, but it is important to ensure that the brine at WIPP, which is different chemically in some respects from the German brine, behaves in a similar fashion with respect to solubility.

HYDROLOGY ABOVE THE SALADO FORMATION

Some scenarios specify that the radionuclide-contaminated brine will be under sufficient pressure to be transported upward into strata above the repository, where it mixes with water of lower salinity, forming a "plume" of contamination.

Much of the radioactivity in the plume may be trapped in the host rock or at least slowed (retarded) by a combination of known effects including precipitation (formation of a solid chemical compound), sorption (chemical reactions with rock surfaces that inhibit transport), and matrix diffusion (egress into fractured rock pores). However, it is difficult to make reliable predictions of these effects. Laboratory experiments with powdered rock are not necessarily good analogues for the natural environment, and even percolation experiments using columns of intact rock cored from the formation are problematical.

Only extensive field experiments can hope to deal directly with the problem of radionuclide transport in the subsurface around WIPP. Unfortunately, the state of art is such that even this approach is not necessarily definitive.

As used in this report, the term "non-Salado hydrology" refers to ground-water flow that occurs in aquifers within the rock formations that lie above the Salado Formation. The Culebra Dolomite is the major aquifer, and it is the only one considered in the 1992 PA model. The water in this aquifer, however, is too salty for consumption by either humans or livestock, making the Culebra a poor potential route for radionuclides to reach the food chain in the vicinity of WIPP.

A considerable amount of field work has been completed to better define the properties of the Culebra, and significant advances have been made. However, the data base and the level of understanding of this flow system still contain notable gaps. A major gap is the lack of attention to a possible radionuclide contamination route through the Dewey Lake Red Beds, in which the water is known to be potable for both humans and livestock. Although a unique definition of the properties of a deep regional system is neither attainable nor necessary, the committee believes uncertainty about some critical issues could and should be reduced further. Field tests in seven new wells drilled at pad H19 at WIPP are scheduled for completion in April 1996, and the results may provide useful new information on contaminant transport and matrix diffusion for the 1996 PA.

Past PA studies have indicated that releases into the Culebra might be serious enough to disqualify WIPP for certification unless it can be shown that source-term concentrations are lower than the high values now conservatively assumed, or that retardation within the Culebra is much greater than zero. In other words, the combination of high plutonium solubility and low retardation under the assumptions of the 1992 PA have the potential to generate noncompliance. A full description of which combinations of parameter values generate noncompliance is an outcome of the PA analysis, and in particular, of the sensitivity analysis derived from the model. Results from past PA studies are illustrative, but use assumptions (such as the number of future boreholes drilled per unit area per unit time) that differ from the parameter assumptions specified by the recent guidance of 40 CFR 194. The future CCDF results of the 1996 PA, especially the

derived sensitivity analysis, would contain the quantitative answer to the dependencies on each parameter of the model.

REPOSITORY DESIGN USING COMPARTMENTATION

Pre-1996 PA calculations were based on a repository design in which separate waste-filled rooms would be connected by the DRZ, which would provide a passage way for gas or liquid. An alternative design would use crushed salt to seal closed each waste-filled room, so that gas or liquid entering one room is prevented from reaching another.

This alternative design provides an approach to demonstrating compliance that the committee expects would lessen the dependencies on (and concerns about) gas generation, the actinide source term, and non-Salado hydrology. Other engineering measures can also be used to improve the repository design. The reconsolidation of salt, discussed earlier, is a powerful design tool, which can be used to create a fully compartmented repository. Access tunnels can be backfilled with some of the salt removed during excavation. Compartmentation and isolation of the waste rooms can be achieved through the use of backfill materials to make effective seals (i.e., barriers to the movement of liquids and gases) to plug all connections made during excavation. Following consolidation of the plugs of backfill material into a tight seal, penetration of any one compartment will have no effect on the rest of the repository.

Such compartmentation would reduce both the probability and the extent of any potential brine flooding associated with human intrusion scenarios. If the rooms were isolated from each other, problems associated with brine entry would be in general reduced, compared with a fully connected repository. By incorporating such a design, the committee believes that the PA model would show that the current state of knowledge about source term and non-Salado hydrology is sufficient to ensure compliance. This expected outcome remains to be demonstrated. However, even should it be possible to certify compliance of WIPP with the available information base, the studies of source term and hydrology should be continued, since these problems are generic to all concepts of geological waste disposal and could provide considerable additional confidence in the long-term performance of WIPP.

SUMMARY, CONCLUSIONS, AND PERSPECTIVE

The readiness of WIPP to open and operate as a radioactive waste repository is the issue at hand in DOE's efforts to demonstrate compliance with EPA regulations. This appendix presents some of the relevant technical issues. The scientific and engineering work of the past several years will be used for the PA computer model calculations by SNL. These calculations compare the projected, estimated performance of WIPP to EPA requirements. The results, a CCDF curve, contain the quantitative assessment of whether the facility can comply with the regulations.

For the undisturbed repository, compliance depends on the ability of the crushed salt seals in the WIPP shafts to develop a sufficiently low permeability to retard the movement of water into and radionuclides out of the repository. This occurs by natural creep processes in a relatively short time (50-100 years), with the salt seals tending to become impermeable eventually, so that releases should be well below the limits of the EPA standard (see Chapter 4).

For a repository disturbed by human intrusion, when evaluated on the basis of reasonable expectation of intrusive activities and their consequences, and using models that would implement available engineering features and do not make overly conservative assumptions, the consensus of the committee is that the WIPP repository could be shown by DOE to comply with the EPA standard.

Compliance with the EPA standard does not address human exposure for the disturbed case. In some release scenarios, such as releases of radionuclides into water at depth in the Salado, the lack of a credible exposure pathway to humans provides an extra margin of reserve in addition to the compliance requirements.

Appendix H

Biographical Sketches of Committee Members

FAIRHURST, Charles, *Chair*—T.W. Bennett Professor of Civil Engineering at the University of Minnesota in Minneapolis. He holds a Ph.D. degree in mining engineering from Sheffield University, England. His professional interests are in rock mechanics, mining, and underground construction, and he has consulted widely in this field for private clients and government agencies—foreign and domestic. Prior to joining the University of Minnesota, he worked in underground mining with the Northwestern Division of the National Coal Board in the U.K. Dr. Fairhurst is a member of the National Academy of Engineering and the Royal Swedish Academy of Engineering Sciences, the American Institute of Mining, Metallurgical & Petroleum Engineers, the American Society of Civil Engineers, and the American Underground Construction Association and was former president of the International Society for Rock Mechanics. He has been a member of the National Research Council's Committee on the Waste Isolation Pilot Plant since 1984.

ADLER, Howard I.—Vice President for Research and Development at Oxyrase, Inc. in Knoxville, Tennessee and adjunct Professor of Microbiology, University of Tennessee. He holds B. S., M. S. and Ph. D. degrees in bacteriology from Cornell University. Dr. Adler has conducted work in this and related fields as senior staff member at Oak Ridge National Laboratory and as Director of Microbiology at Oak Ridge Associate Universities. He is a member of the American Society of Microbiology, American Association for the Advancement of Science, and Sigma Xi. He has been a member of the National Research Council's Committee on the Waste Isolation Pilot Plant since 1991.

BLOMEKE, John 0.—Holds B. S. and M. S. degrees in chemical engineering from the University of Texas and a Ph. D. degree in chemical engineering from Georgia Tech. He cultivated his professional interest in chemical engineering while working on the Manhattan Project at the University of Chicago and the Clinton Laboratory in Oak Ridge, Tennessee. Dr. Blomeke has been actively engaged at Oak Ridge National Laboratory in the development and evaluation of commercial nuclear fuel waste management. He has remained very active in this field through his retirement in 1988. He has been a member of the National Research Council's Committee on the Waste Isolation Pilot Plant since 1980.

CLARK, Sue B.—Assistant Professor of Chemistry at Washington State University in Pullman, Washington. She holds a B.S. degree from Lander College and M. S. and Ph.D. degrees in inorganic/radiochemistry from Florida State University. Prior to joining Washington State University, she worked as an assistant research ecologist at the Savannah River Ecology Laboratory at the University of Georgia and a research assistant at Katholieke University te Leuven in Belgium. Dr. Clark has received the Westinghouse Savannah River Company's Total Quality Achievement Award and the National Academy of Science's Young Investigator Award. She is a member of Sigma Xi and the American Chemical Society. She has been a member of the National Research Council's Committee on the Waste Isolation Pilot Plant since 1994.

ERNSBERGER, Fred M.—Adjunct Professor of Materials Science at the University of Florida, Gainesville, Florida. He holds an A.B. degree in chemistry from Ohio Northern University and a Ph.D. degree in physical chemistry from Ohio State University. Prior to joining the University of Florida he was Senior Scientist in the Glass Research Center for PPG Industries, Inc. He has received the George W. Morey Award, the Toledo Glass & Ceramic Award, and the Bleininger Medal of the American Ceramic Society. Dr. Ernsberger is a member of the American Chemical Society and a fellow of the American Ceramic Society. He has been a member of the National Research Council's Committee on the Waste Isolation Pilot Plant since 1978.

EWING, Rodney C.—Regents Professor in the Department of Earth and Planetary Sciences at the University of New Mexico where he has been a member of the faculty for 22 years. He holds a B.S. degree in geology from Texas Christian University (summa cum laude) and M.S. and Ph.D. degrees in geology from Stanford University. His professional interests are in mineralogy and materials science. He has conducted research in Sweden, Germany, Australia, and Japan as well as the United States. Dr. Ewing is a fellow of the Geological Society of America and the Mineralogical Society of America. Presently, he is the vice-president and president-elect of the International Union of Materials Research Societies. He has served on the National Research Council's the Waste Isolation Pilot Plant Committee since 1984.

GARRICK, B. John—Chairman and founder of PLG, Inc., an international engineering, applied science, and management consulting firm in Newport Beach, California. He received his B.S. degree from Brigham Young University and his M.S. and Ph.D. degrees in engineering and applied science from the University of California, Los Angeles. His professional interests involve risk assessment in fields such as nuclear energy, space and defense, chemical and petroleum, and transportation. He received the Society for Risk Analysis Distinguished Achievement Award and was appointed to the U.S. Nuclear Regulatory Commission's Advisory Committee on Nuclear Waste in 1994, of which he is now Vice Chairman. Dr. Garrick is a member of the National Academy of Engineering and is currently Vice Chairman of the National Research Council's Board on Radioactive Waste Management. He has been a member of the National Research Council's Committee on the Waste Isolation Pilot Plant since 1989.

KONIKOW, Leonard F.—Research Hydrologist with the U.S. Geological Survey in Reston, Virginia. He holds a B.A. degree in geology from Hofstra University and M.S. and Ph.D. degrees from Pennsylvania State University. His research interests include the development and application of solute-transport models to ground-water contamination problems. Previously, he has served on the Editorial Board of *Ground Water* journal, as Associate Editor for *Water Resources Research*, and as Chairman of the Hydrogeology Division of the Geological Society of America. He served on the National Research Council's Committee on Ground Water Modeling Assessment and has been a member of the Committee on the Waste Isolation Pilot Plant since 1989.

KRAUSKOPF, Konrad B.—Professor of Geochemistry Emeritus at Stanford University in California. He holds an undergraduate degree from the University of Wisconsin, and Ph.D. degrees in chemistry from the University of California and in geology from Stanford University. He has received several honors, including Day Medal from the Geological Society of America and the Goldschmidt Medal from the Geochemical Society. He is a member of the National Academy of Sciences and has served as chairman of the National Research Council's Board on Radioactive Waste Management. He has been a member of the National Research Council's Committee on the Waste Isolation Pilot Plant since 1978.

ROY, Della M.—Professor of Materials Science Emerita at the Materials Research Laboratory and Materials Science and Engineering at Pennsylvania State University. She holds a B.S. degree from the University of Oregon and M. S. and Ph.D. degrees from Pennsylvania State University. Her research interests include material synthesis and characterization in inorganic, ceramic, cement, and mineral systems, and nuclear and chemical waste management. She is founding editor and editor-in-chief of *Cement and Concrete Research*.

She is a fellow of the American Ceramic Society, the American Concrete Institute, the Mineralogical Society of America, and the American Association for the Advancement of Science. Dr. Roy is a member of the National Academy of Engineering. She has served on the National Research Council's Committee on the Waste Isolation Pilot Plant since 1995.

WAITE, David A.—Vice President of Health and Safety at Kaiser Hill Co., LLC in Rocky Flats, Colorado. He holds a B.A. degree in physics from Emporia State University, an M.S. degree in radiation physics from Vanderbilt University, and a Ph.D. in general engineering from Oklahoma State University. Prior to joining Kaiser Hill, he conducted work in this and related fields with other companies, including Envirosphere and the Battelle Memorial Institute. Dr. Waite is a member of Lambda Delta Lambda, Sigma Pi Sigma, and the American Board of Health Physics. He has been a member of the National Research Council's Committee on the Waste Isolation Pilot Plant since 1994.

WHIPPLE, Chris G.—Vice President of ICF Kaiser Engineers in Oakland, California. He holds a B.S. degree from Purdue University and a Ph.D. degree in engineering science from the California Institute of Technology. His professional interests are in risk assessment, and he has consulted widely in this field for private clients and government agencies. Prior to joining ICF Kaiser Engineers, he conducted work in this and related fields at the Electric Power Research Institute. He served on the National Research Council's Board on Radioactive Waste Management from 1985 to 1995 and as its Chair from 1992 to March 1995. He has been a member of the National Research Council's Committee on the Waste Isolation Pilot Plant since 1989.

ZORDAN, Thomas A.—President of Zordan Associates, Inc. and concurrently, Senior Project Manager at ICF Kaiser Engineers in Pittsburgh, Pennsylvania. He holds a B.S. degree from Northern Illinois University and a Ph.D. degree in physical chemistry from the University of Louisville. His professional and research interests include environmental management, risk assessment, waste management, software development, and thermochemistry. Prior to joining ICF Kaiser he conducted work in this and related fields with companies such as Science Applications International Corporation, Westinghouse Electric Corporation, and Gulf Research & Development Company. He has served on the National Research Council's Committee on the Waste Isolation Pilot Plant since 1992.

Appendix I

Glossary

absorbed dose—energy imparted to matter in a volume element by ionizing radiation, divided by the mass of irradiated material in that volume element. The SI derived unit of absorbed dose is the gray (Gy); 1 Gy = 100 rad = 1 J per kilogram.

accessible environment—region outside a rectangular block of rock 5 km long on each side placed symmetrically around the Waste Isolation Pilot Plant repository, with vertical sides extending upward to the land surface and indefinitely downward into the earth.

actinide—element with atomic number 90 (thorium) or greater.

actinide source term—cumulative amounts, concentrations in solution, and chemical nature of all radioactive materials to be disposed of at WIPP, which could be moved in solution or suspension (or retarded by sorption on surrounding rock units) from the WIPP repository into the environment.

advection—solute-transport process in which chemical species are carried through an aquifer by the flowing ground water in which they are dissolved.

anhydrite—anhydrous calcium sulfate.

aquifer—formation, group of formations, or part of a formation that contains sufficient saturated permeable material to yield significant quantities of water to wells and springs.

Becquerel—radiation unit equal to one disintegration per second.

Castile Formation—oldest of the Ochoan sequence of rocks, consisting of alternating layers of anhydrite and thin layers of limestone, with several thick layers of halite. At depth, highly pressurized pockets of brine have been found in this formation.

clastic—term describing rocks or sediments made up of fragments of other rocks. For example, sandstone and shale are clastic rocks.

colloid—particles of size ranging from 10^{-3} to 1 micron (10^{-9} to 10^{-6} m), finer than clay size, and held in suspension in a medium such as water. Because of their fine size, colloids have very large surface areas.

committed effective dose—a measure of human exposure to radioactivity. An ingestion of, inhalation of, or direct contact with radionuclides deposits energy from radioactive decays into bodily tissues. This absorbed dose is multiplied by a weighting factor dependent on the type of radiation to calculate an equivalent dose. The weighted sum of equivalent doses to all bodily tissues and summed over the time of exposure is the committed effective dose, measured in rem. The SI unit, the Sievert (Sv), is equivalent to 100 rem.

compartmentation—filling of tunnels and drifts around rooms and panels, so that the waste is isolated in compartments.

complex—stable association of a metal ion with one or more bulky groups (ligands), resulting in a profound change in the properties of the ion.

corrensite—clay mineral with a structure that represents interstratification of vermiculite and chlorite structures.

creep—slow movement over time of salt as shear stresses cause movement within or between individual crystals.

Culebra Dolomite—second-oldest member of the Rustler Formation ranging from approximately 7-8 m thick at the WIPP site. The Culebra consists of dolomite with some clay minerals. Because it is a water-bearing unit, the Culebra is important to the ground–water flow model for the WIPP site.

curie—measure of the quantity of radioactive material in a sample, equal to 3.7×10^{10} disintegrations per second.

Darcy's law—empirical hydraulic principle stating that the flow rate of a fluid through a porous material is proportional to the hydraulic gradient and the hydraulic conductivity of the material.

Darcy velocity—volumetric flow rate per unit surface area of sample, on the assumption that flow is laminar and that inertia can be neglected. This is more properly called the "specific discharge."

Delaware Basin—sedimentary basin in which the WIPP site is located. The Delaware Basin formed in a Permian sea and was gradually filled with thick, extensive layers of sediments and evaporite deposits.

Dewey Lake Red Beds—youngest of the Ochoan series of rocks, consisting of thin beds of clay, silt, and sandy sediments, red in color.

diffusion length—distance over which chemical species diffuse, or spread out and thoroughly mix, over a period of time, approximated as the square root of the product of diffusivity and time.

distribution coefficient—ratio of the amount of a substance sorbed to the amount in solution when chemical equilibrium exists (also see isotherm); ratio of the concentrations of a substance in two phases at equilibrium.

disturbed repository performance—radionuclide releases from the repository as a result of reasonably foreseeable natural processes as well as inadvertent human intrusion.

disturbed rock zone (DRZ)—localized region of permeability consisting of microfractures in the salt. The DRZ is generated by stress concentrations at a new surface and spreads at a slow and decreasing rate from that surface.

dolomite—a sedimentary rock consisting mostly of the mineral dolomite, calcium magnesium carbonate.

effective porosity—fraction of interconnected void space per unit volume of porous material. It is less than the total porosity because it includes only pores that comprise active flow paths through the medium and excludes voids in isolated or dead-end pores.

Eh—measure of the oxidation-reduction state of a solution.

electrical resistivity—characteristic of a solution depending on the concentration of ions in the solution and measured as the resistance of a centimeter cube of a substance to the passage of a current perpendicular to two parallel faces of the cube.

EPA unit— isotopic-specific unit of measure, used by Sandia National Laboratories, that is equal to the number of curies represented by the release limit of the EPA requirements in 40 CFR part 191.

evaporite—natural deposit of water-soluble salts formed by the evaporation of a body of water.

flocculation—aggregation or clumping together of particles suspended in solution so that they precipitate.

flow path—pathway a particle of ground water would travel, starting from a specified point.

halite—common rock salt, with chemical formula NaCl.

half-life—time required for 50 percent of any amount of a radioactive element to decay.

high-level waste—radioactive waste resulting from the reprocessing of spent fuel rods from nuclear reactors, or of other radioactive materials used for defense purposes.

human intrusion—inadvertent creation of a radiation pathway by human actions.

hydraulic conductivity—capacity for a porous medium to conduct water; defined formally as the volume of water at the existing kinematic viscosity that will move in unit time under a unit hydraulic gradient through a unit area measured at right angles to the direction of flow through a porous material. This coefficient incorporates the properties of both the porous medium and the liquid.

hydraulic diffusivity—K/S_s, where K is hydraulic conductivity and S_s is specific storage. It represents the hydraulic conductivity (K) of a saturated porous medium when the unit volume of water diffusing is that involved in changing the head of a unit amount in a unit volume of medium. The time involved for a given head change to occur at a particular point in response to a greater change in head at another point is inversely proportional to the hydraulic diffusivity.

hydraulic gradient—change in static (or hydraulic) head per unit of distance in a given direction. If not specified, the direction generally is understood to be that of the maximum rate of decrease in head.

hydraulic head—height above a standard datum of the surface of a column of water that can be supported by the static pressure at a given point.

The hydraulic head is the sum of the elevation head and the pressure head. Under conditions to which Darcy's law may be applied, the velocity of ground-water flow is so low that the velocity head is negligible. Stated more simply, hydraulic head is the height of free surface of water above a given subsurface point.

hydrostatic pressure—pressure exerted in all directions by a body of water at rest.

hydrostratigraphic unit—formation, part of a formation, or group of formations of considerable lateral extent that compose the geologic framework for a reasonably distinct hydrologic system.

ionic strength—half the sum of the products of molality and square of charge for each ion in a solution.

ionization—any process by which an atom, molecule, or ion gains or loses electrons.

ionizing radiation—radiation consisting of directly ionizing particles.

isotherm—line or curve describing the relation between the amount of a substance sorbed and the amount in solution when the chemical system is in equilibrium. The relation is typically very sensitive to temperature changes, so data used to construct such a curve are normally based on experiments conducted at a constant temperature—hence, the name isotherm. The slope of a linear isotherm is often called a distribution coefficient.

isotropic—equal in all directions.

kinetics—general term for rates of reaction.

langbeinite—potash mineral with chemical formula $K_2Mg_2(SO_4)_3$ of economic importance as a source of K_2SO_4.

ligand—compound or molecule in solution that can combine with the ion or molecule of an actinide to form a complex.

lithostatic pressure—pressure exerted by a column of overlying rock at a point in the earth's crust.

Magenta Dolomite Member—second-youngest member of the Rustler Formation, 7-8.5 m thick, consisting of dolomite with gypsum and bearing ground water.

marker bed—horizontally extensive nonhalite interbed in the Salado, numbered from the top of the Salado to the bottom and used to keep the repository at the same level within the Salado.

matrix diffusion—transfer of solute mass by diffusion, driven by a concentration gradient, between an actively flowing part of a ground–water system and low-permeability (often stagnant) parts of the saturated medium. An example is diffusion from a high-permeability fracture into dead-end pores of the adjacent rock matrix.

McNutt Member—a 120-m-thick series of beds midway between the top and bottom of the Salado Formation, which lies stratigraphically above the WIPP repository. The McNutt member contains economically significant potash deposits that have been mined in the area for many years.

molality—number of moles of solute per 1,000 g of solvent in a solution.

Nash Draw—broad shallow valley that developed as a stream channel at the end of the Pleistocene (1.6 million to 10,000 years ago) glaciation. Rocks of the Rustler Formation are exposed in the walls of the Nash Draw.

numerical dispersion—artificial smearing of calculated gradients in a dependent variable (such as concentration) that results from discretization in a numerical solution to the governing partial differential equations (such as the solute-transport equation).

Ochoan—Late Permian stratigraphic series consisting of, from lowermost to uppermost, the Castile, Salado, Rustler, and Dewey Lake Formations.

Ostwald's step rule—rule stating that an unstable phase passes through successive steps of increasing stability as it becomes a final stable phase.

parameter—an algebraic symbol representative of a well defined physical quantity with a numerical value. An adjustable parameter is envisioned to assume any value within its range (between the maximum and minimum numerical bounds). Any particular choice of a value renders a parameter a numerical constant.

Pennsylvanian Period—period of time from 320 million to 286 million years ago.

permeability—capacity of a material to transmit fluids. A measure of the relative ease with which a porous medium can transmit a liquid under a potential gradient. Permeability depends on the size, shape, and degree of interconnectedness of pores and is generally measured in millidarcies. It is a property of the medium alone and independent of the nature of the liquid.

performance assessment—risk-based assessment of the safety performance of a nuclear waste facility.

Permian Period—period of time from 286 million to 245 million years ago[1].

pH—measure of the acidity of a solution phase; negative logarithm of the hydrogen ion concentration.

---

[1]Older geologic dating extended the Permian to 225 Ma; more recent methods date the end as 245 Ma.

Pitzer parameters—correction factors for calculating solubilities in strong electrolyte solutions.

potash—group of potassium salts, mined for various uses including plant fertilizers.

potentiometric elevation—elevation that represents the hydraulic head. For an aquifer, it is defined by the level to which water will rise in a tightly cased well. The water table is a particular potentiometric surface.

radiolysis—decomposition brought about by high-energy irradiation.

radionuclide—radioactive nuclide.

recharge—addition of water to the saturated zone; either natural or artificial.

redox potential—number specifying the "oxidation state" of a solution or, alternatively stated, a measure of the ability of a natural environment to control the equilibrium of oxidation-reduction reactions. Redox potential is a synonym for oxidation potential.

retardation factor—in solute-transport analysis, a parameter that describes the ratio of the net apparent velocity of the concentration of a particular chemical species to the velocity of a nonreactive species (or of water). It is proportional to the slope of a sorption isotherm; thus, if the isotherm is nonlinear, the retardation factor is not constant and depends on solute concentration.

risk—probability of a selected set of consequences from the existence of a hazard (e.g., nuclear waste in a repository) or the operation of a facility.

Rustler Formation—second youngest Ochoan formation, overlying the Salado Formation, and consisting of five sequences (members) of thin-bedded strata. The lowermost beds consist of mudstone and sandstone interbedded with evaporites. The upper part of the formation consists of alternating evaporite and dolomite beds. The Culebra Dolomite Member is the second member from the bottom of the formation. The total thickness of the Rustler Formation near the WIPP site is approximately 100 m.

safety—protection against the adverse health effects of hazards (such as radiation and chemicals), regardless of cause.

Salado Formation—second-oldest Ochoan geologic formation consisting of a 230-million year old deposit of rock salt (halite) in near-horizontal beds, in total 200-400 m thick. Very thin layers of clay, anhydrite, and potash minerals are interbedded with the halite beds. Lying at a depth of approximately 660 m (2,160 ft) at the WIPP site, the Salado hosts the WIPP repository.

saturated zone—that part of the water-bearing material in which all voids are filled with water under pressure greater than atmospheric.

sorbing tracer tests—field experiments designed to measure the amount of sorption of a substance.

sorption—chemical reaction processes that result in the accumulation of ions and molecules at a fluid-solid interface. Sorption depends on the balance between the affinity of a substance for a surface and its affinity for an aqueous solution. The term is sometimes used in a general sense to include ion exchange, surface complexation, and surface precipitation.

spalling/spallation—breaking off of very thin sheets from a rock surface.

specific discharge—rate of discharge of ground water per unit area measured at right angles to the direction of flow (see also Darcy velocity).

specific storage—volume of water released from or taken into storage per unit volume of a porous medium per unit change in head. This property is normally considered in problems of three-dimensional transient flow in a compressible ground–water system.

specific weight—mass per unit volume of a substance.

storativity (or storage coefficient)—volume of water an aquifer releases from or takes into storage per unit surface area of the aquifer per unit change in head.

sylvite—potash mineral with chemical formula KCl, the principal ore of potassium.

tortuosity—ratio of the length of a fluid particle's path to the length of a straight line between the beginning and ending points of the path. According to this definition, the tortuosity will always have a value greater than one. Note that tortuosity is sometimes defined on the basis of an inverse of this relation; in such cases it will always have a value less than one.

transmissivity—rate at which water of prevailing properties is transmitted through a unit width of an aquifer under a unit hydraulic gradient. For horizontal flow in a homogeneous aquifer, it is equal to the hydraulic conductivity times the saturated thickness. Transmissivity is measured in units of length squared per time (e.g., m^2/s).

transuranic waste—radioactive waste consisting of radionuclides with atomic numbers greater than 92 in excess of agreed limits. A more precise definition, in DOE Order 5820.2A, EPA regulation 40 CFR 191, and the Land Withdrawal Act, is waste that is not high level waste that is "contaminated with alpha-emitting radionuclides of atomic number greater than 92 and half-lives greater than 20 years in concentrations greater than 100 nanocuries per gram." The regulatory definition excludes actinide elements with atomic numbers between 90 and 92 (most significantly, Th and U isotopes), which is in agreement with the literal meaning of "transuranic." However, common usage of "transuranic waste" is often understood to include all actinides.

undisturbed repository performance—radionuclide releases from the repository as a result of reasonably foreseeable natural processes

unsaturated zone—subsurface zone above the water table in which some of the pore spaces are occupied by air and some by water.

waste characterization—process of identifying and classifying the chemical, physical, and radiological constituents of each drum of waste.

waterflooding—technique used in the secondary recovery of petroleum and gas whereby water is injected into a petroleum or gas reservoir so that the pressure of the water expels the petroleum or gas.

water table—surface in an unconfined ground–water system at which the pressure is atmospheric, in the field environment defined by the levels at which water stands in wells that penetrate the saturated zone just far enough to hold standing water.

Appendix J

List of Acronyms and Symbols

ε	effective porosity
AEA	Atomic Energy Act
AEC	Atomic Energy Commission
AST	actinide source term
BRAGFLO	computer code modeling regional hydrology
CAO	U.S. Department of Energy Carlsbad Area Office
c.c.c.	critical coagulation concentration
CCDF	complementary cumulative distribution function
CH	contact handled
Ci	curie
DCCA	Draft Compliance Certification Application
DEIS	Draft Environmental Impact Statement
DOE	U.S. Department of Energy
DRZ	disturbed rock zone
E1	a scenario in which the WIPP repository is breached by one borehole extending down to a Castille brine pocket
E1E2	a scenario in which the WIPP repository is breached by two boreholes, one extending down to a Castille brine pocket and sealed above the repository horizon, and the other providing a flow path above the Salado
EACBS	Engineered Alternatives Cost/Benefit Study
EAMP	Engineered Alternatives Multidisciplinary Panel
EATF	Engineered Alternatives Task Force
EEG	State of New Mexico Environmental Evaluation Group
EPA	U.S. Environmental Protection Agency
ERDA	Energy Research and Development Administration
FEPs	features, events, and processes
GI	gastrointestinal

HI	human intrusion
HYDROCOIN	International Hydrologic Code Intercomparison Project
INTRAVAL	International Project to Study Validation of Geosphere Transport Models
K_d	distribution coefficient
LANL	Los Alamos National Laboratory
LWA	Land Withdrawal Act
Ma	millions of years ago
MB	marker bed
MPa	megapascal
MTHM	metric tons of heavy metal
NAS	National Academy of Sciences
NMED	New Mexico Environment Department
NRC	National Research Council
ONWI	Office of Nuclear Waste Isolation
PA	performance assessment
QA	quality assurance
RCRA	Resource Conservation and Recovery Act
R_f	retardation factor
RH	Remote handled
SECO	computer code used in modeling hydrology
SIT	specific-ion interaction theory
SNL	Sandia National Laboratories
SPM	systems prioritization method
STTP	Source Term Test Program
TRU	literally, transuranic (i.e., with atomic number greater than that of uranium); however, TRU waste is commonly used to refer to all actinide-bearing waste (i.e., containing thorium and uranium isotopes)
VOC	volatile organic compound
WIPP	Waste Isolation Pilot Plant

Appendix K

Bibliography

40 CFR 191. 1995. Environmental radiation protection standards for management and disposal of spent nuclear fuel, high-level and transuranic radioactive wastes. Code of Federal Regulations Title 40, Pt. 191.

40 CFR 194. 1996. Criteria for the certification and recertification of the Waste Isolation Pilot Plant's compliance with the 40 CFR Part 191 disposal regulations; final rule. Federal Register 61(28) (February 9):5224-5245.

40 CFR 268. 1995. Land disposal restrictions. Code of Federal Regulations Title 40, Pt. 268.

Ahlbom, K., J. Andersson, P. Andersson, T. Ittner, C. Ljunggren, and S. Tirén. 1992. Finnsjön Study Site. Scope of Activities and Main Results: SKB Technical Report 92-33. Stockholm: Swedish Nuclear Fuel and Waste Management Company.

Ahrens, E. H. 1995. Data Report on the Small-Scale Seal Performance Tests, Series F Grouting Experiment in Room L3. SAND93-1000. Albuquerque, N.Mex.: Sandia National Laboratories.

Ahrens, E. H., and F. D. Hansen. 1995. Large-Scale Dynamic Compaction Demonstration Using WIPP Salt: Fielding and Preliminary Results. SAND95-1941. Albuquerque, N.Mex.: Sandia National Laboratories.

Anderson, M. P., and W. W. Woessner. 1992. Applied Groundwater Modeling: Simulation of Flow and Advective Transport. San Diego: Academic Press.

Barker, J. M., and G. S. Austin. 1995. Overview of the Carlsbad Potash District. Pp. III-1–III-26 in Evaluation of Mineral Resources at the Waste Isolation Pilot Plant (WIPP) Site, Vol. 2. Socorro, N.Mex.: New Mexico Bureau of Mines and Mineral Resources, Campus Station.

Barth, F. E. 1982. Prehistoric saltmining at Hallstatt. Bulletin, Institute of Archaeology, London 19:31-43.

Beauheim, R. L. 1987. Interpretations of Single-Well Hydraulic Tests Conducted at and near the Waste Isolation Pilot Plant (WIPP) Site, 1983-1987. SAND87-0039. Albuquerque, N.Mex.: Sandia National Laboratories.

Beauheim, R. L. 1989. Interpretation of H-11b4 Hydraulic Tests and the H-11 Multipad Pumping Test of the Culebra Dolomite at the Waste Isolation Pilot Plant (WIPP) Site. SAND89-0536. Albuquerque, N.Mex.: Sandia National Laboratories.

Beauheim, R. L., G. J. Saulnier, Jr., and J. Avis. 1991. Interpretation of Brine-Permeability Tests of the Salado Formation at the Waste Isolation Pilot Plant Site: First Interim Report. SAND90-0083. Albuquerque, N.Mex.: Sandia National Laboratories.

Beauheim, R. L., R. M. Roberts, T. F. Dale, M. D. Fort, and W. A. Stensrud. 1993. Hydraulic Testing of Salado Formation Evaporites at the Waste Isolation Pilot Plant: Second Interpretive Report. SAND92-0533. Albuquerque, N.Mex.: Sandia National Laboratories.

Beauheim, R. L., W. W. Wawersik, and R. M. Roberts. 1993. Coupled permeability and hydrofracture tests to assess the waste-containment properties of fractured anhydrite. International Journal of Rock Mechanics and Mining Sciences & Geomechanics Abstracts 30, 7: 1159-1163.

Beauheim, R. L., S. M. Howarth, P. Vaughn, S. W. Webb, and K.W. Larson. 1995. Integrating Modeling and Experimental Programs to Predict Brine and Gas Flow at the Waste Isolation Pilot Plant. SAND94-0599A and CONF-941053-1. Albuquerque, N.Mex.: Sandia National Laboratories.

Bechtel National, Inc. 1979. Waste Isolation Pilot Plant: Title I Design Report, Vols. 1 and 1a. Nuclear Fuel Operations, San Francisco, Calif.

Bein, A., S. D. Hovorka, R. S. Fisher, and E. Roedder. 1991. Fluid inclusions in bedded Permian halite, Palo Duro Basin, Texas: evidence for modification of seawater in evaporite brine-pools and subsequent early diagenesis. Journal of Sedimentary Petrology 61(1):1-14.

Berest, P., and B. Brouard. 1996. Behavior of sealed solution–mined caverns. Paper presented at SALT-IV— Fourth Conference on the Mechanical Behavior of Salt, Montreal, Canada, June 17-18. (Proceedings to be published in early 1997. Clausthal-Zellerfeld, Germany: Trans Tech Publications.)

Borns, D. J. 1995. Implications of geophysical surveys in the WIPP underground on the interpretation of the relative roles of the three proposed conceptual models for Salado fluid flow. Memo October 3, 1994. Appendix I in Systems Prioritization Method—Iteration 2. Baseline Position Paper: Salado Formation Fluid Flow and Transport Contaminant Group. S. Howarth et al., eds. Sandia National Laboratories.

Bottrell, S. H., ed. 1996. Proceedings of the Fourth International Symposium on the Geochemistry of the Earth's Surface. Leeds, U.K.: University of Leeds.

Brady, B. H. G. and E. T. Brown. 1987. Rock Mechanics for Underground Mining. London: Routledge, Chapman and Hall.

Bredehoeft, J. D. 1988. Will salt repositories be dry? EOS (Transactions American Geophysical Union) 69(9):121-131.

Bredehoeft, J. D., and L. F. Konikow. 1993. Reply to comment by G. de Marsily, P. Combes, and P. Goblet. Advances in Water Resources 15(6):371-372.

Brinster, K. F. 1991. Preliminary Geohydrologic, Conceptual Model of the Los Medaños Region Near the Waste Isolation Pilot Plant for the Purpose of Performance Assessment. SAND89-7147. Albuquerque, N.Mex.: Sandia National Laboratories.

Broadhead, R. F., F. Luo, and S. W. Speer. 1995. Oil and Gas Resource Estimates. Pp. XI-1–XI-135 in Evaluation of Mineral Resources at the Waste Isolation Pilot Plant (WIPP) Site, Vol. 3. Socorro, N.Mex.: New Mexico Bureau of Mines and Mineral Resources, Campus Station.

Brodsky, N. S. 1990. Crack Closure and Healing Studies in WIPP Salt Using Compressional Wave Velocity and Attenuation Measurements: Test Methods and Results. SAND90-7076. Albuquerque, N.Mex.: Sandia National Laboratories.

Brodsky, N. S. 1995. Thermodynamical Damage Recovery Parameters for Rocksalt from the Waste Isolation Pilot Plant. SAND93-7111. Albuquerque, N.Mex.: Sandia National Laboratories.

Brodsky, N. S., F. D. Hansen, and T. W. Pfeifle. 1996. Properties of Dynamically Compacted WIPP Salt. SAND96-0838C. Albuquerque, N Mex.: Sandia National Laboratories. Paper presented at SALT IV— Fourth Conference on the Mechanical Behavior of Salt, Montreal, Canada, June 17-18. (Proceedings to be published in early 1997. Clausthal-Zellerfeld, Germany: Trans Tech Publications.)

Brush, L. H. 1990. Test Plan for Laboratory and Modeling Studies of Repository and Radionuclide Chemistry for the Waste Isolation Pilot Plant. SAND90-0266. Albuquerque, N.Mex.: Sandia National Laboratories.

Brush, L. H. 1994. Position Paper on Gas Generation in the Waste Isolation Pilot Plant. Draft for Environmental Protection Agency and Stakeholder Review. Albuquerque, N.Mex.

Brusseau, M. L. 1994. Transport of reactive contaminants in heterogeneous porous media. Reviews of Geophysics 32(3):285-313.

Butcher, B. M. 1990. Preliminary Evaluation of Potential Engineered Modifications for the Waste Isolation Pilot Plant (WIPP). SAND89-3095. Albuquerque, N.Mex.: Sandia National Laboratories.

Butcher, B. M., and F. T. Mendenhall. 1993. A Summary of the Models Used for the Mechanical Response of Disposal Rooms in the Waste Isolation Pilot Plant with Regard to Compliance with 40 CFR 191, Subpart B. SAND92-0427. Albuquerque, N.Mex.: Sandia National Laboratories.

Butcher, B. M., C. F. Novak, and M. Jercinovic. 1991. The Advantages of a Salt/Bentonite Backfill for Waste Isolation Pilot Plant Disposal Rooms. SAND90-3074. Albuquerque, N.Mex.: Sandia National Laboratories.

Callahan, G. D., M. C. Loken, L. D. Hurtado, and F. D. Hansen. 1996. Evaluations of constitutive models for crushed salt. Paper presented at SALT-IV—Fourth Conference on the Mechanical Behavior of Salt. June 17-18, Montreal, Canada. (Proceedings to be published in early 1997. Clausthal-Zellerfeld, Germany: Trans Tech Publications.)

Center for Nuclear Waste Regulatory Analyses. 1994. Background Report on the Use and Elicitation of Expert Judgment. CNWRA 94-019. Prepared for U.S. Nuclear Regulatory Commission.

Chapman, J. B. 1988. Chemical and Radiochemical Characteristics of Groundwater in the Culebra Dolomite, Southeastern New Mexico. EEG-39. Albuquerque, N.Mex.: Environmental Evaluation Group.

Churchill, R. V. 1972. Operational Mathematics. McGraw-Hill Publishing Company.

Churchill, R. V., and J. W. Brown. 1987. Fourier Series and Boundary Value Problems. McGraw-Hill, Inc.

Cranwell, R. M., R. V. Guzowski, J. E. Campbell, and N. R. Ortiz. 1990. Risk Methodology for Geologic Disposal of Radioactive Waste: Scenario Selection Procedure SAND80-1429. Sandia National Laboratories, prepared for U.S. Nuclear Regulatory Commission, NUREG/CR-1667.

Dale, T., and L. D. Hurtado. 1996. WIPP Air-Intake Shaft Disturbed Rock Zone Study. Paper presented at SALT-IV—Fourth Conference on the Mechanical Behavior of Salt, Montreal, Canada, June 17-18. (Proceedings to be published in early 1997. Clausthal-Zellerfeld, Germany: Trans Tech Publications.)

Davies, P. B. 1989. Variable-Density Ground Water Flow and Paleohydrology in the Waste Isolation Pilot Plant (WIPP) Region, Southeastern New Mexico. U.S. Geological Survey Open-File Report 88-490. Albuquerque, N.Mex.: U. S. Geological Survey.

Davies, P. B., J. F. Pickens, and R. L. Hunter. 1991. Complexity in the Validation of Ground Water Travel Time in Fractured Flow and Transport Systems. SAND89-2379. Albuquerque, N.Mex.: Sandia National Laboratories.

De Marsily, G., P. Combes, and P. Goblet. 1992. Comment on "Ground water models cannot be validated" by L. F. Konikow and J.D. Bredehoeft. Advances in Water Resources 15(6):371-372.

Deal, D. E., R. J. Abitz, D. S. Belski, J. B. Clark, M. E. Crawley, and M. L. Martin. 1991. Brine Sampling and Evaluation Program 1989 Report. Carlsbad, N.Mex.: Westinghouse Electric Corporation, Waste Isolation Division.

Deal, D. E., R. J. Abitz, J. Myers, J. B. Case, D. S. Belski, M. L. Martin, and W. M. Roggenthem. 1991. Brine Sampling and Evaluation Program 1990 Report. Carlsbad, N.Mex.: Westinghouse Electric Company, Waste Isolation Division.

Detournay, E. and A. H.-D. Cheng. 1988. Poroelastic response of a borehole in a non-hydrostatic stress field. International Journal of Rock Mechanics and Mining Sciences and Geomechanics Abstracts 25(3):171-182.

DOE (U.S. Department of Energy). 1980-1983. Waste Isolation Pilot Plant Safety Analysis Report. WIPP SAR (includes Amendments 1-6). Albuquerque Operations Office, N.Mex.

DOE. 1990. Draft Environmental Impact Statement (DEIS): Waste Isolation Pilot Plant, Vols. 1 and 2. DOE/EIS-0026-D. Washington, D.C.

DOE. 1990. Final Supplement, Environmental Impact Statement, Waste Isolation Pilot Plant. DOE/EIS-0026-FS. Office of Environmental Restoration and Waste Management, Washington, D.C.

DOE. 1991. Evaluation of the Effectiveness and Feasibility of the Waste Isolation Pilot Plant Engineered Alternatives: Final Report of the Engineered Alternatives Task Force, Vol. I and II. DOE/WIPP 91-007, Revision 0. Carlsbad, N.Mex.: WIPP Project Office.

DOE. 1995. Draft Title 40 CFR 191 Compliance Certification Application for the Waste Isolation Pilot Plant. DRAFT-DOE/CAO-2056. March 31 and update, July 31. Carlsbad, N.Mex.: Carlsbad Area Office.

DOE. 1995. Engineered Alternatives Cost/Benefit Study Final Report. DOE/WIPP 95-2135, Revision 0. Carlsbad, N.Mex.: Carlsbad Area Office.

DOE. 1995. Transuranic Waste Baseline Inventory Report. Revision 2. DOE/CAO 95-1121, 3 vols. Carlsbad, N.Mex.: Carlsbad Area Office.

DOE. 1995. Waste Isolation Pilot Plant Sealing System Design Report. DOE/WIPP-95-3117.

EPA (U.S. Environmental Protection Agency), Office of Radiation Programs. 1988. Limiting Values of Radionuclide Intake and Air Concentration and Dose Conversion Factors for Inhalation, Submersion, and Ingestion. 1988 Federal Guidance Report No. 11, EPA-520/1-88-020. Washington, D.C.

EPA, Office of Radiation and Indoor Air. 1996. Response to Comments. 40 CFR 194: Criteria for the Certification and Re-Certification of the Waste Isolation Pilot Plant's Compliance with the 40 CFR Part 191 Disposal Regulations. EPA 402-R-96-001.

Eyermann, T. J., L. L. Van Sambeek, and F. D. Hansen. 1995. Case Studies of Sealing Methods and Material Used in the Salt and Potash Mining Industries. SAND95-1120. Albuquerque, N.Mex.: Sandia National Laboratories.

Felmy A. R. and J. H. Weare. 1986. The prediction of borate mineral equilibria in natural waters: application to Searles Lake, California. Geochimica et Cosmochimica Acta 50:2771-2783.

Finley, R. E., and J. R. Tillerson. 1992. WIPP Small Scale Seal Performance Tests—Status and Impacts. SAND91-2247. Albuquerque, N.Mex.: Sandia National Laboratories.

Francis, A. J., and J. B. Gillow. 1994. Effects of Microbial Processes on Gas Generation Under Expected WIPP Repository Conditions. Progress Report Through 1992. SAND93-7036. Albuquerque, New Mex.: Sandia National Laboratories.

Franke, O. L., T. E. Reilly, and G. D. Bennett. 1987. Definition of boundary and initial conditions in the analysis of saturated ground water flow systems—An introduction. Ch. B5 in Techniques of Water-Resources Investigations of the United States Geological Survey, Book 3: Applications of Hydraulics. U.S. Geological Survey.

Freeze, G. A., and T. L. Christian-Frear. In preparation. Modeling Brine Inflow to Room Q. A Numerical Investigation of Flow Mechanics.

Freeze, G. A., K. W. Larson, and P. B. Davies. 1995. Coupled Multiphase Flow and Closure Analysis of Repository Response to Waste-Generated Gas at the Waste Isolation Pilot Plant (WIPP). SAND93-1986. Sandia National Laboratories. Albuquerque, N.Mex.

Freeze, G. A., K. W. Larson, and P. B. Davies. 1995. A Summary of Methods for Approximating Salt Creep and Disposal Room Closure in Numerical Models of Multiphase Flow. SAND94-0251. Albuquerque, N.Mex.: Sandia National Laboratories.

Garabedian, S. P., D. R. LeBlanc, L. W. Gelhar, and M. A. Celia. 1991. Large-scale natural gradient tracer test in sand and gravel, Cape Cod, Massachusetts: 2, analysis of tracer moments for a nonreactive tracer. Water Resources Research 27(5):911-924.

Garven, G. 1995. Continental-scale groundwater flow and geologic processes. Annual Review of Earth and Planetary Sciences 23:89-117.

Gelhar, L. W., C. Welty, and K. R. Rehfeldt. 1992. A critical review of data on field-scale dispersion in aquifers. Water Resources Research 28(7):1955-1974.

Giammattei, V. M., and N. G. Reichert. 1975. The Mimbres Classic Black-on-White: Art of a Vanished Race. Calistoga, Cal.: Dillon Tyler.

Goode, D. J., and L. F. Konikow. 1989. Modification of a Method-of-Characteristics Solute-Transport Model to Incorporate Decay and Equilibrium-Controlled Sorption or Ion Exchange. U.S. Geological Survey Water-Resources Investigations Report 89-4030.

Grenthe, I., and H. Wanner. 1992. Guidelines for the Extrapolation to Zero Ionic Strength. NEA-TBD-2, revision 2. Gif-sur-Yvette, France: Nuclear Energy Agency, Organization for Economic Co-operation and Development.

Griswold, G. B. 1995. Future Mining Technology Pp. IV-1–IV-5 in Evaluation of Mineral Resources at the Waste Isolation Pilot Plant (WIPP) Site. Vol. 2. Socorro, N.Mex.: New Mexico Bureau of Mines and Mineral Resources, Campus Station.

Gulick, C. W., Jr., J. A. Boa, Jr., D. M. Walley, and A. D. Buck. 1980. Borehole Plugging Materials Development Program, Report 2. SAND79-1514. Albuquerque, N.Mex.: Sandia National Laboratories.

Hansen, F. D., and E. H. Ahrens. 1996. Large-Scale Dynamic Compaction of Natural Salt. SAND96-0792C. Albuquerque, N.Mex.: Sandia National Laboratories. Paper presented at SALT IV—Fourth Conference on the Mechanical Behavior of Salt, Montreal, Canada, June 17-18. (Proceedings to be published in early 1997. Clausthal-Zellerfeld, Germany: Trans Tech Publications.)

Hareland, G. 1995. The Past-Decade Developments and Future Trends in Oil-Well Drilling, Completion, and Stimulation, with Special Applications to Developments at the WIPP Site. Chapter X in Evaluation of Mineral Resources at the Waste Isolation Pilot Plant (WIPP) Site, Vol. 3. Socorro, N.Mex.: New Mexico Bureau of Mines and Mineral Resources, Campus Station.

Harvie C. E., N. Møller, and J. H. Weare. 1984. The Prediction of Mineral Solubilities in Natural Waters: The Na-K-Mg-Ca-H-Cl-SO_4-OH-HCO_3-CO_3-CO_2-H_2O system to high ionic strengths at 25°C. Geochimica et Cosmochimica Acta 48:723-751.

Helton, J. C., J. W. Garner, R. D. McCurley, and D. K. Rudeen. 1991. Sensitivity Analysis Techniques and Results for Performance Assessment at the Waste Isolation Pilot Plant. SAND90-7103. Albuquerque, N.Mex.: Sandia National Laboratories.

Helton, J. C., J. W. Garner, R. P. Rechard, D. K. Rudeen, and P. N. Swift. 1992. Preliminary Comparison with 40 CFR Part 191, Subpart B for the Waste Isolation Pilot Plant, December 1991. Vol. 4: Uncertainty and Sensitivity Analysis Results. SAND91-0893/4. Albuquerque, N.Mex. Prepared by Sandia National Laboratories for the U. S. Department of Energy.

Howarth, S., K. Larson, T. Christian-Frear, R. Beauheim, D. Borns, D. Deal, A. L. Jensen, K. Knowles, D. Powers, R. Roberts, M. Tierney, and S. Webb. 1995. Systems Prioritization Method—Iteration 2. Baseline Position Paper: Salado Formation Fluid Flow and Transport Contaminant Group. Albuquerque, N.Mex.: Sandia National Laboratories.

Ibaraki, M., and E. A. Sudicky. 1995. Colloid-facilitated contaminant transport in discretely fractured porous media. 1. Numerical formulation and sensitivity analysis. Water Resources Research 31: 2945-2960.

ICRP (International Commission on Radiological Protection). 1985. ICRP Publication 46: Radiation protection principles for the disposal of solid radioactive waste. Annals of the ICRP. Vol. 15, no. 4. Elmsford, N.Y.: Pergamon Press.

ICRP. 1990. ICRP Publication 60: 1990 Recommendations of the International Commission on Radiological Protection, Annals of the ICRP. Vol. 21, no. 1-3. Elmsford, N.Y.: Pergamon Press.

Jensen, A. L., R. L. Jones, E. N. Lorusso, and C. L. Howard. 1993. Large-Scale Brine Inflow Data Report for Room Q Prior to November 25, 1991. SAND92-1173. Albuquerque, N.Mex.: Sandia National Laboratories.

Jones, B. F., and S. K. Anderholm. 1993. Normative analysis of brines from the Salado Formation and underlying strata, SE New Mexico. EOS (Transactions of the American Geophysical Union) 74(16):326. Abstract.

Jones, B. F., and S. K. Anderholm. 1996. Some geochemical considerations of brines associated with bedded salt repositories. Pp. 343-353 in Proceedings of the Fourth International Symposium on the Geochemistry of the Earth's Surface, S. H. Bottrell, ed. Leeds, U.K.: University of Leeds.

Kaplan, S., and B. J. Garrick. 1981. On the quantitative definition of risk. Risk Analysis 1(1):11-27.

Kay, M., and E. H. Colbert. 1965. Stratigraphy and Life History. New York: John Wiley and Sons.

Kelley, V. A., and G. J. Saulnier, Jr. 1990. Core Analyses for Selected Samples from the Culebra Dolomite at the Waste Isolation Pilot Plant Site. SAND90-7011. Albuquerque, N.Mex.: Sandia National Laboratories.

Kim, J. I. 1994. Actinide colloids in natural aquifer systems. Materials Research Society Bulletin XIX(12):47-53.

Konikow, L. F., and J. D. Bredehoeft. 1992. Ground water models cannot be validated. Advances in Water Resources 15(1):75-83.

Kushner, D. J. 1978. Life in High Salt and Solute Concentrations—Halophilic Bacteria. Pp. 317-335 in Microbial Life in Extreme Environments, D. J. Kushner, ed. New York: Academic Press.

Laaksoharju, M., C. Degueldre, and C. Skårman. 1995. Studies of Colloids and Their Importance for Repository Performance Assessment. SKB Technical Report 95-24. Stockholm: Swedish Nuclear Fuel and Waste Management Company.

Lappin, A. R. 1988. Summary of Site-Characterization Studies Conducted from 1983 Through 1987 at the Waste Isolation Pilot Plant (WIPP) Site, Southeastern New Mexico. SAND88-0157. Albuquerque, N.Mex.: Sandia National Laboratories.

Lappin, A. R., and R. L. Hunter, eds. 1989. Systems Analysis, Long-Term Radionuclide Transport, and Dose Assessments, Waste Isolation Pilot Plant (WIPP), Southeastern New Mexico. SAND89-0462. Albuquerque, N.Mex.: Sandia National Laboratories.

LaVenue, A. M., A. Haug, and V. A. Kelley. 1988. Numerical Simulation of Ground Water Flow in the Culebra Dolomite at the Waste Isolation Pilot Plant (WIPP) Site. Second Interim Report. SAND88-7002. Albuquerque, N.Mex.: Sandia National Laboratories.

LaVenue, A. M., T. L. Cauffman, and J. F. Pickens. 1990. Ground Water Flow Modeling of the Culebra Dolomite, Vol. 1: Model Calibration. SAND89-7068/1. Albuquerque, N.Mex.: Sandia National Laboratories.

LaVenue, A. M., B. S. RamaRao, G. de Marsily, and M. G. Marietta. 1995. Pilot point methodology for automated calibration of an ensemble of conditionally simulated transmissivity fields, 2. Application. Water Resources Research 31(3):495-516.

Lienert, C., S. A. Short, and H. R. von Gunten. 1994. Uranium Infiltration from a River to Shallow Groundwater. Geochimica et Cosmochimica Acta 58(24):5455-5463.

Marietta, M. C., S. G. Bertram-Howery, D. R. Anderson, K. F. Brinster, R. V. Guzowski, H. Lussolino, and R. P. Rechard. 1989. Performance Assessment Methodology Demonstration: Methodology Development for Evaluation Compliance with EPA 40 CFR 191, Subpart B, for the Waste Isolation Pilot Plant. SAND89-2027. Albuquerque, N.Mex.: Sandia National Laboratories.

Marshal, K. 1976. Interfaces in Microbial Ecology. Cambridge, Mass.: Harvard University Press.

McKinley, I. G., and D. Savage. 1993. Comparison of Solubility Databases Used for HLW Performance Assessment. Radiochimica Acta 66/67:657-665.

McTigue, D. F. 1991. Horizontal Darcy Flow to Room Q. Memo June 24. Appendix E in Systems Prioritization Method—Iteration 2. Baseline Position Paper: Salado Formation Fluid Flow and Transport Contaminant Group, S. Howarth et al., eds. Albuquerque, N.Mex.: Sandia National Laboratories.

McTigue, D. F. 1993. Permeability and Hydraulic Diffusivity of Waste Isolation Pilot Plant Repository Salt Inferred from Small-Scale Brine Inflow Experiments. SAND92-1911. Albuquerque, N.Mex.: Sandia National Laboratories.

McTigue, D. F. 1995. Calculation of brine flux and cumulative brine volume for room Q, based on a Darcy-flow model. Memo April 3, 1989. Pp. E-35–E-36 in Systems Prioritization Method—Iteration 2. Baseline Position Paper: Salado Formation Fluid Flow and Transport Contaminant Group, S. Howarth et al., eds. Albuquerque, N.Mex.: Sandia National Laboratories.

McTigue, D. F. 1995. A model for brine inflow due to salt "damage." Memo August 2, 1990. Pp. E-3–E-29 in Systems Prioritization Method Iteration 2. Baseline Position Paper: Salado Formation Fluid Flow and Transport Contaminant Group, S. Howarth et al., eds. Albuquerque, N.Mex.: Sandia National Laboratories.

Mercer, J. W. 1987. Compilation of Hydrologic Data from Drilling the Salado and Castile Formations Near the Waste Isolation Pilot Plant (WIPP) Site in Southeastern New Mexico. SAND86-0954. Albuquerque, N.Mex.: Sandia National Laboratories.

Meyer, D., and J. J. Howard, eds. 1983. Evaluation of Clays and Clay Minerals for Application to Repository Sealing. Prepared for the Office of Nuclear Waste Isolation (ONWI), Columbus, Ohio, by D'Appolonia and Pennsylvania State University.

Miller, C. W., and L. V. Benson. 1983. Simulation of solute transport in a chemically reactive heterogeneous system: model development and application. Water Resources Research 19(2):381-391.

Moench, A. F. 1984. Double-porosity models for a fissured groundwater reservoir with fracture skin. Water Resources Research (20):834-846.

Molecke, M. A. 1979. Gas Generation from Transuranic Waste Degradation: Data Summary and Interpretation. SAND79-1245. Sandia National Laboratories, Albuquerque, N.Mex.

Molecke, M. A. 197b. Gas generation potential from TRU wastes. Pp. 3-1–3-21 in Summary of Research and Development Activities in Support of Waste Acceptance Criteria for WIPP. SAND79-1305. Sandia National Laboratories. Albuquerque, N.Mex.

Munson, D. E. In press. Constitutive Model of Creep in Rock Salt Applied to Underground Room Closure. International Journal of Rock Mechanics and Mining Sciences and Geomechanics Abstracts.

Munson, D. E., K. L. DeVries, D. M. Schiermeister, W. F. DeYonge, and R. L. Jones. 1992. Measured and calculated closures of open and brine filled shafts and deep vertical boreholes in salt. Pp. 439-448 in Rock Mechanics Proceedings of the 33rd U.S. Symposium, , Sweeney Convention Center, Santa Fe, New Mexico, 3-5 June 1992, J. R. Tillerson and W. R. Wawersik, eds. Rotterdam; Brookfield: A. A. Balkema. (Also published as SAND91-1869. Albuquerque N.Mex.: Sandia National Laboratories.)

Munson, D. E., D. J. Borns, M. K. Pickens, D. J. Holcomb, and S. E. Bigger. 1995. Systems Prioritization Method—Iteration 2 Baseline Position Paper: Rock Mechanics: Creep, Fracture, and Disturbed Rock Zone (DRZ). Albuquerque, N.Mex.: Sandia National Laboratories.

NEA (Nuclear Energy Agency). 1984. Long-Term Radiation Protection Objectives for Radioactive Waste Disposal. Paris, France: Nuclear Energy Agency, Organization for Economic Co-operation and Development.

Neill, R. H., Lokesh C., W. W.-L. Lee, T. M. Clemo, M. K. Silva, J. W. Kenney, W. T. Bartlett, and B. A. Walker. 1996. Review of the WIPP Draft Application to Show Compliance with EPA Transuranic Waste Disposal Standards. EEG-61. Environmental Evaluation Group, N.Mex.

New Mexico Bureau of Mines and Mineral Resources, Campus Station. 1995. Evaluation of Mineral Resources at the Waste Isolation Pilot Plant (WIPP) Site. Socorro, N.Mex. 4 vols.

Nichols, M. D., and E. P. Laws. 1995. Letters to Senators Larry E. Craig and Dirk Kempthorne. U.S. Environmental Protection Agency.

Nitsche, H., K. Roberts, R. Xi, T. Prussin, K. Becraft, I. Al Mahamid, H. B. Silver, S. A. Carpenter, R. C. Gatti, and C. F. Novak. 1994. Long term plutonium solubility and speciation studies in a synthetic brine. Radiochimica Acta 66/67: 3-8.

Novak, C. F. 1992. An Evaluation of Radionuclide Batch Sorption Data on Culebra Dolomite for Aqueous Compositions Relevant to the Human Intrusion Scenario for the Waste Isolation Pilot Plant. SAND91-1299. Albuquerque, N.Mex.: Sandia National Laboratories.

Novak, C. F. 1995. Actinide Chemistry Research Supporting the Waste Isolation Pilot Plant (WIPP): FY94 Results. SAND94-2274. Albuquerque, N.Mex.: Sandia National Laboratories.

Novak , C. F. 1995. The Waste Isolation Pilot Plant (WIPP) Actinide Source Term: Test Plan for the Conceptual Model and the Dissolved Concentration Submodel. SAND95-1985. Albuquerque, N.Mex.: Sandia National Laboratories.

Novak, C. F., and K. E. Roberts. 1995. Thermodynamic Modeling of Neptunium (V) Solubility in $Na-CO_3-HCO_3-Cl-ClO_4-H-OH-H_2O$ Electrolytes. Scientific Basis for Nuclear Waste Management XVIII. Proceedings of the Materials Research Society, 353:1119-1128.

Novak, C. F., R. F. Weimer, H. W. Papenguth, Y. K. Behl, D. A Lucero, F. Gelbard, and J. A. Romero. Unpublished. Actinide Source Term and Chemical Retardation Programs letter and briefing materials on solubility, colloid, and chemical retardation work, distributed at Actinide Source Term and Chemical Retardation Programs meeting. June 29, 1995, Albuquerque, N.Mex.

NRC (National Research Council). 1957. The Disposal of Radioactive Waste on Land. Washington, D.C.: National Academy Press.

NRC. 1970. Disposal of Solid Radioactive Wastes in Bedded Salt Deposits. Washington, D.C.: National Academy Press.

NRC. 1979. WIPP Letter Report to Mr. Sheldon Meyers, Program Director of the U.S. DOE Office of Nuclear Waste Management. Washington, D.C.: National Academy Press.

NRC. 1979. WIPP Letter Report to Mr. Sheldon Meyers, Program Director of the U.S. DOE Office of Nuclear Waste Management. Washington, D.C.: National Academy Press.

NRC. 1983. A Study of the Isolation System for Geologic Disposal of Radioactive Waste. Washington, D.C.: National Academy Press.

NRC. 1984. Review of the Scientific and Technical Criteria for the Waste Isolation Pilot Plant. Washington, D.C.: National Academy Press.

NRC. 1987. WIPP Letter Report to Mr. John Mathur, U.S. DOE Office of Defense Waste and Transportation Management. Washington, D.C.: National Academy Press.

NRC. 1988. WIPP Letter Report to the Honorable John S. Herrington, Secretary of the Department of Energy. Washington, D.C.: National Academy Press.

NRC. 1988. WIPP Letter Report to Mr. Critz George, U.S. DOE Office of Defense Waste and Transportation Management. Washington, D.C.: National Academy Press.

NRC. 1989. WIPP Letter Report. Washington, D.C.: National Academy Press.

NRC. 1990. Rethinking High Level Radioactive Waste. Washington, D.C.: National Academy Press.

NRC. 1991. WIPP Letter Report to Mr. Leo Duffy, Director of the U.S. DOE Office of Environmental Restoration and Waste Management. Washington, D.C.: National Academy Press.

NRC. 1992. WIPP Letter Report (addressing the underground experimental plan with TRU wastes) to the Honorable Leo P. Duffy, Assistant Secretary of the U.S. DOE Office of Environmental Restoration and Waste Management. Washington, D.C.: National Academy Press.

NRC. 1995. Technical Bases for Yucca Mountain Standards. Washington, D.C.: National Academy Press.

Papenguth, H. W. and Y. K. Behl. 1996. Test Plan for Evaluation of Colloid-Facilitated Actinide Transport at the WIPP. Test Plan 96-01. Albuquerque, N.Mex.: Sandia National Laboratories.

Papenguth, H. W., and Y. K. Behl. 1996. Test Plan for Evaluation of Dissolved Actinide Retardation at the WIPP. Test Plan 96-02. Albuquerque, N.Mex.: Sandia National Laboratories.

Pickard, Lowe and Garrick, Inc.; Westinghouse Electric Corporation; and Fauske and Associates, Inc. 1981. Zion Probabilistic Safety Study. Prepared for Commonwealth Edison Company.

Pitzer, K. S. 1973. Thermodynamics of electrolytes. I. Theoretical basis and general equations. J. Phys. Chem. 77: 268-277.

Pitzer, K. S. 1977. Electrolyte theory--Improvements Since Debye and Hückel. Acct. Chem. Res., 10: 371-377.

Pitzer, K. S., ed. 1991. Activity Coefficients in Electrolyte Solutions. 2nd ed. Boca Raton, Fla.: CRC Press.

Powers, D. 1996. Tracing Early Breccia Pipe Studies, Waste Isolation Pilot Plant, Southeastern New Mexico: A Study of the Documentation Available and Decision-Making During the Early Years of WIPP. SAND94-0991. Albuquerque, N.Mex.: Sandia National Laboratories.

Powers, D. W., S. J. Lambert, S.-E. Shaffer, L. R. Hill, and W. D. Weart, eds. 1978. Geological Characterization Report, Waste Isolation Pilot Plant (WIPP) Site, Southeastern New Mexico, Vols. I and II. SAND78-1596. Albuquerque, N.Mex.: Sandia National Laboratories.

Prindle, N. H., F. T. Mendenhall, D. M. Boak, W. Beyeler, D. Rudeen, R. C. Lincoln, K. Trauth, D. R. Anderson, M. G. Marietta, and J. C. Helton. 1996. The Second Iteration of the Systems Prioritization Method: A Systems Prioritization and Decision-Aiding Tool for the Waste Isolation Pilot Plant. 3 vols. SAND95-2017. Albuquerque, N.Mex.: Sandia National Laboratories.

Reardon, E. J. 1981. Kd's—Can they be used to describe reversible ion sorption reactions in contaminant migration? Ground Water 19(3):279-286.

Rechard, R. P. 1995. An Introduction to the Mechanics of Performance Assessment Using Examples of Calculations Done for the Waste Isolation Pilot Plant Between 1990 and 1992. SAND93-1378. Albuquerque, N. Mex.: Sandia National Laboratories.

Rechard, R. P., P. J. Roache, R. L. Blaine, A. P. Gilkey, and D. K. Rudeen. 1991. Quality Assurance Procedures for Computer Software Supporting Performance Assessments of the Waste Isolation Pilot Plant. SAND90-1240. Albuquerque, N.Mex.: Sandia National Laboratories.

Rechard, R. P., D. K. Rudeen, and P. J. Roache. 1992. Quality Assurance Procedures for Analyses and Report Reviews Supporting Performance Assessments of the Waste Isolation Pilot Plant. SAND91-0428. Albuquerque, N.Mex.: Sandia National Laboratories.

Rechard, R. P., K. M. Trauth, and R. V. Guzowski. 1992. Quality Assurance Procedures for Parameter Selection and Use of Expert Judgment Panels Supporting Performance Assessments of the Waste Isolation Pilot Plant. SAND91-0429. Albuquerque, N.Mex.: Sandia National Laboratories.

Reed, D. T., S. Okajima, and M. K. Richmann. 1994. Stability of Plutonium(VI) in Selected WIPP Brines. Radiochimica Acta 66/67:95-101.

Reilly, T. E., O. L. Franke, H. T. Buxton, and G. D. Bennett. 1987. A Conceptual Framework for Ground Water Solute-Transport Studies with Emphasis on Physical Mechanisms of Solute Movement. U.S. Geological Survey Water-Resources Investigation Report 87-4191.

RE/SPEC. Unpublished. Calculations done under contract to Sandia National Laboratories WIPP Shaft Sealing Program. RSI Calculation File 325/11/03.

RE/SPEC. Unpublished. Calculations done under contract to Sandia National Laboratories WIPP Shaft Sealing Program. RSI Calculation File 325/11/04.

RE/SPEC. Unpublished. Calculations done under contract to Sandia National Laboratories WIPP Shaft Sealing Program. RSI Calculation File 325/11/05.

RE/SPEC. 1995. External Memorandum to L. Diane Hurtado, Sandia National Laboratories, November 10. RSI(RCO)-325/11/95/23.

Roddy, J. W., H. C. Claiborne, R. C. Ashline, P. J. Johson, and B. T. Rhyne. 1986. Physical and Decay Characteristics of Commercial LWR Spent Fuel. ORNL/TM-9591/V16R1. Oak Ridge, Tenn.: Oak Ridge National Laboratory.

Roedder, E., W. M. d'Angelo, A. F. Dorrzapf, Jr., and P. J. Aruscavage. 1987. Composition of Fluid Inclusions in Permian Salt Beds, Palo Duro Basin, Texas, U.S.A. Chemical Geology 61:79-90.

Rossmanith, H.-P., ed. 1995. Mechanics of Jointed and Faulted Rock: Proceedings of the 2nd International Conference on the Mechanics of Jointed and Faulted Rock, MJFR-2, Vienna, Austria, 10-14 April, 1995. Rotterdam: A. A. Balkema.

Roy, D. M., M. W. Grutzeck, and L. D. Wakeley. 1983. Selection and Durability of Seal Materials for a Bedded Salt Repository, Preliminary Studies. ONW1-479. Prepared for Office of Nuclear Waste Isolation, Battelle Memorial Institute, Columbus, Ohio, by Pennsylvania State University.

Roy, D. M., M. W. Grutzeck, and L. D. Wakeley. 1985. Salt Repository Seal Materials. A Synopsis of Early Cementitious Materials Development, BMI/ONWI-536. Office of Nuclear Waste Isolation, Battelle Memorial Institute, Columbus, Ohio, by Pennsylvania State University.

Sandia National Laboratories. 1979. Summary of Research and Development Activities in Support of Waste Acceptance Criteria for WIPP. SAND79-1305. Albuquerque, N.Mex.

Sandia National Laboratories. 1991. Preliminary Comparison with 40 CFR Part 191. Subpart B for the Waste Isolation Pilot Plant. December 1991, Vol. 3: Reference Data. SAND91-0893/3. Albuquerque, N.Mex.

Sandia National Laboratories. 1992. Preliminary Performance Assessment for the Waste Isolation Pilot Plant, December 1992, Vols. 1-5. SAND92-0700. Albuquerque, N.Mex.

Sandia National Laboratories. 1995. SPM-2 Report, Revision 1, Vols. I and II. Albuquerque, N.Mex.

Sewards, T., R. Glenn, and K. Keil. 1991. Mineralogy of the Rustler Formation in the WIPP-19 Core. SAND87-7036. Albuquerque, N.Mex.: Sandia National Laboratories.

Sexton, T. J. 1996. Memo, April 11, to Jan van Schilfgaarde, Agricultural Research Service, U.S. Department of Agriculture.

Sharp, J. M., Jr., N. I. Robinson, R. C. Smyth-Boulton, and K. L. Milliken. 1995. Fracture Skin Effects in Groundwater Transport. Pp. 449-454 in Mechanics of Jointed and Faulted Rock, H.-P. Rossmanith, ed. Rotterdam: A. A. Balkema.

Shoesmith, D. W., S. Sunder, M. G. Bailey, and N. H. Miller. 1996. Corrosion of used nuclear fuel in aqueous perchlorate and carbonate solutions. Journal of Nuclear Materials 227:287-299.

Siegel, M. D., and S. Anderholm. 1994. Geochemical evolution of groundwater in the Culebra Dolomite near the Waste Isolation Pilot Plant, southeastern New Mexico, USA. Geochimica et Cosmochimica Acta 58(10):2299-2323.

Siegel, M. D., S. J. Lambert, and K. L. Robinson, eds. 1991. Hydrogeochemical Studies of the Rustler Formation and Related Rocks in the Waste Isolation Pilot Plant Area, Southeastern New Mexico. SAND88-0196. Albuquerque, N.Mex.: Sandia National Laboratories.

Silva, M. K. 1994. Implications of the Presence of Petroleum Resources on the Integrity of the WIPP. EEG-55; DOE/AL/58309-55. Albuquerque, N.Mex.: Environmental Evaluation Group.

Silva, M. K. 1996. Fluid Injection for Salt Water Disposal and Enhanced Oil Recovery as a Potential Problem for the WIPP: Proceedings of a June 1995 Workshop and Analysis. EEG-62. Albuquerque, N.Mex.: Environmental Evaluation Group.

Stoelzel, D. M. 1996. Current Petroleum Practices and Their Application to WIPP Area Development. Pp. 79-91 in Fluid Injection for Salt Water Disposal and Enhanced Oil Recovery as a Potential Problem for the WIPP: Proceedings of a June 1995 Workshop and Analysis, M. K. Silva, ed. Albuquerque, N.Mex: Environmental Evaluation Group.

Stormont, J. C. 1984. Plugging and Sealing Program for the Waste Isolation Pilot Plant (WIPP). SAND84-1057. Albuquerque, N.Mex.: Sandia National Laboratories.

Stormont, J. C., ed. 1986. Development and Implementation: Test Series A of the Small-Scale Seal Performance Tests. SAND85-2602. Albuquerque, N.Mex.: Sandia National Laboratories.

Stormont, J. C. 1988. Preliminary Seal Design Evaluation for the Waste Isolation Pilot Plant. SAND87-3083. Albuquerque, N.Mex.: Sandia National Laboratories.

Stormont, J. C., C. L. Howard, and J. J. K. Daemen. 1991. In Situ Measurements of Rock Salt Permeability Changes Due to a Nearby Excavation. SAND90-3134. Albuquerque, N.Mex.: Sandia National Laboratories.

Swedish Nuclear Fuel and Waste Management Company. 1995. Feasibility Study for Siting of a Deep Repository Within the Storuman Municipality: SKB Technical Report 95-08, Stockholm.

Swedish Nuclear Power Inspectorate. 1987. The International HYDROCOIN Project—Background and Results. Paris, France: Organization for Economic Co-operation and Development.

Swedish Nuclear Power Inspectorate. 1990. The International INTRAVAL Project—Background and Results. Paris, France: Organization for Economic Co-operation and Development.

Tanji, K. K., and B. Yaron, eds. 1994. Management of Water Use in Agriculture. New York: Springer-Verlag.

Tillerson, J. R., and W. R. Wawersik, eds. 1992. Rock Mechanics Proceedings of the 33rd U.S. Symposium, , Sweeney Convention Center, Santa Fe, New Mexico, 3-5 June 1992. Rotterdam; Brookfield: A. A. Balkema.

U.S. Congress. 1980. The U.S. Department of Energy National Security and Military Applications of Nuclear Energy Authorization Act of 1980. Public Law 96-164. 93 Stat. 1259. Enacted December 1979. 96th Congress.

U.S. Congress. 1992. Waste Isolation Pilot Plant Land Withdrawal Act P. L. 102-579. Legislative Report for the 102nd Congress.

U.S. Congress, House of Representatives. 1995. Waste Isolation Pilot Plant Land Withdrawal Amendment Act. H.R. 1663. Introduced May 17, 1995 by Representative Skeen (R-N.Mex.). 104th Congress, 1st Session.

U.S. Congress, Senate. 1995. Waste Isolation Pilot Plant Land Withdrawal Amendment Act. S1402. 104th Congress, 1st Session.

U.S. Congress, Senate. 1996. Waste Isolation Pilot Plant Land Withdrawal Amendment Act. Amendment No. 4085 to S1745. 104th Congress. C. R. pp. S6587-S6591.

U.S. Geological Survey. 1987. Techniques of Water-Resources Investigations of the United States Geological Survey, Book 3: Application of Hydraulics.

U.S. Nuclear Regulatory Commission. 1975. Reactor Safety Study: An Assessment of Accident Risks in U.S. Commercial Nuclear Power Plants. WASH-1400, NUREG-75/014.

Valocchi, A. J. 1984. Describing the transport of ion-exchanging contaminants using an effective K_d approach. Water Resources Research 20(4):499-503.

Van der Leeden, F., L. A. Cerrillo, and D. W. Miller. 1975. Ground water pollution problems in the northwestern United States. EPA-660/3-75-018. U.S. Environmental Protection Agency.

Van Sambeek, L. L., D. D. Luo, M. S. Lin, W. Ostrowski, and D. Oyenuga. 1993. Seal Design Alternatives Study. SAND92-7340. Albuquerque, N.Mex.: Sandia National Laboratories.

Wakeley, L. D., P. T. Harrington, and C. A. Weiss, Jr. 1993. Properties of Salt-Saturated Concrete and Grout after Six Years n Situ at the Waste Isolation Pilot Plant. SAND93-7019. Albuquerque, N.Mex.: Sandia National Laboratories.

Wakeley, L. D., T. S. Poole, and J. P. Burkes. 1994. Durability of Concrete Materials in High Magnesium Brine. SAND93-7073. Albuquerque, N.Mex.: Sandia National Laboratories.

Wakeley, L. D., P. T. Harrington, and F. D. Hansen. 1995. Variability in Properties of Salado Mass Concrete. SAND94-1495. Albuquerque, N.Mex.: Sandia National Laboratories.

Waste Isolation Plant Land Withdrawal Amendments Act. 1996. National Defense Authorization Act for Fiscal Year 1997, Subtitle F. (P.L. 104-201, September 23).

Webb, S. W., and K. W. Larson. 1996. The Effect of Stratigraphic Dip on Brine Inflow and Gas Migration at the Waste Isolation Pilot Plant. SAND 94-0932. Albuquerque, N.Mex.: Sandia National Laboratories.

Wikberg, P., G. Gustafson, I. Rhén, and R. Stanfors. 1991. Äspö Hard Rock Laboratory. Evaluation and Conceptual Modeling Based on the Pre-investigations 1986-1990. SKB Technical Report 91-22. Stockholm: Swedish Nuclear Fuel and Waste Management Company.

Williams, N. 1995. The trials and tribulations of cracking the prehistoric code. Science 269:923-924.

Yaron, B., and H. Frenkel. 1994. Water Suitability for Agriculture. In Management of Water Use in Agriculture, K. K. Tanji and B. Yaron, eds. New York: Springer-Verlag.